Six-Minute Solutions

for Structural Engineering (SE) Exam Morning Breadth Problems

Third Edition

Christine A. Subasic, PE

Professional Publications, Inc. • Belmont, California

Benefit by Registering This Book with PPI

- Get book updates and corrections.
- Hear the latest exam news.
- Obtain exclusive exam tips and strategies.
- Receive special discounts.

Register your book at **www.ppi2pass.com/register**.

Report Errors and View Corrections for This Book

PPI is grateful to every reader who notifies us of a possible error. Your feedback allows us to improve the quality and accuracy of our products. You can report errata and view corrections at **www.ppi2pass.com/errata**.

SIX-MINUTE SOLUTIONS FOR STRUCTURAL ENGINEERING (SE) EXAM MORNING BREADTH PROBLEMS
Third Edition

Current printing of this edition: 1

Printing History

edition number	printing number	update
2	2	Minor corrections.
2	3	Minor corrections. Updated to *AASHTO LRFD Bridge Design Specifications*, 4th ed.
3	1	New edition. Code update. Exam specifications update. Copyright update.

Copyright © 2012 by Professional Publications, Inc. (PPI). All rights reserved. No part of this publication may be reproduced, stored in a retrieval system, or transmitted, in any form or by any means, electronic, mechanical, photocopying, recording, or otherwise, without the prior written permission of the publisher.

Printed in the United States of America.

PPI
1250 Fifth Avenue, Belmont, CA 94002
(650) 593-9119
www.ppi2pass.com

ISBN: 978-1-59126-391-3

Library of Congress Control Number: 2012944831

Table of Contents

ABOUT THE AUTHOR iv

PREFACE, DEDICATION, AND ACKNOWLEDGMENTS v

INTRODUCTION
   Exam Format vii
   This Book's Organization vii
   How to Use This Book vii

CODES AND REFERENCES xi

NOMENCLATURE xiii

VERTICAL BREADTH PROBLEMS 1

LATERAL BREADTH PROBLEMS 15

SOLUTIONS VERTICAL BREADTH PROBLEMS 23

SOLUTIONS LATERAL BREADTH PROBLEMS 61

About the Author

Christine A. Subasic, PE, is a consulting architectural engineer licensed in North Carolina and Virginia. Ms. Subasic graduated with a bachelor of architectural engineering degree, structures option, with honors and high distinction, from The Pennsylvania State University. For more than 20 years, she has specialized in structural and masonry design and standards development. Ms. Subasic has design experience in both commercial and residential construction. Her clients include commercial businesses as well as several trade associations.

Ms. Subasic is an active member of ASTM International, an organization committed to the development of construction industry standards. Ms. Subasic has served on the board of directors for The Masonry Society, and she is currently active on their Design Practices Committee, Architectural Practices Committee, and Sustainability Committee.

Ms. Subasic is the author of numerous articles and technical publications. She has authored and reviewed chapters in the *Masonry Designers' Guide*, published by The Masonry Society, and co-authored *An Investigation of the Effects of Hurricane Opal on Masonry*. She has served as a subject matter expert for revisions of timber and masonry material in PPI publications. Her articles on many aspects of masonry have appeared in the magazines *Masonry Construction*, *Masonry*, and *STRUCTURE*.

Ms. Subasic is also active at home, with her three boys, at their school, and in her church community.

Preface, Dedication, and Acknowledgments

The Structural Engineering (SE) exam, administered by the National Council of Examiners for Engineering and Surveying (NCEES), is created from problems developed by educators and professional engineers representing consulting, government, and industry. The SE exam is designed to test examinees' understanding of both conceptual and practical engineering concepts. Problems from past exams are not available from NCEES or any other source. However, NCEES does identify the general subject areas covered on the exam.

The SE exam is broken into two eight-hour components: vertical forces and lateral forces. Each component is further subdivided into a four-hour morning breadth exam and a four-hour afternoon depth exam. Both breadth exams contain 40 multiple-choice problems that each take an average of six minutes to solve and cover a variety of structural engineering topics related to buildings and bridges. The depth exams each comprise four one-hour essay problems covering either buildings or bridges topics. Examinees must choose to take either the buildings or bridges module and must take the same module for both the vertical and lateral components. This book only contains morning breadth problems for the vertical and lateral components.

The topics covered in this third edition of *Six-Minute Solutions for Structural Engineering (SE) Exam Morning Breadth Problems* have been updated to coincide with those subject areas identified by NCEES for the SE breadth exams. Included among these problem topics are analysis of structures, design and details of structures, and construction administration. Problems and solutions have also been updated to reflect current NCEES SE design standards.

The problems presented in this book are representative of the type and difficulty of problems you will encounter on the SE breadth exams. The problems are both conceptual and practical, and they are written to provide varying levels of difficulty. Though you probably won't encounter problems on the breadth exams exactly like those presented here, working these problems and reviewing the solutions will increase your familiarity with the breadth exam problems' form, content, and solution methods. This preparation will help you considerably during the exam.

Problems and solutions have been carefully prepared and reviewed to ensure that they are appropriate and understandable and that they were solved correctly. If you find errors or discover an alternative, more efficient way to solve a problem, please bring it to PPI's attention so your suggestions can be incorporated into future editions. You can report errors and keep up with the changes made to this book by logging on to PPI's errata website at **www.ppi2pass.com/errata**.

I would like to dedicate this book to Shawn, my husband and biggest supporter, without whom I could never have worked all the hours necessary to complete this project. I would also like to acknowledge the support of my friends and mentors in the industry, particularly Phillip Samblanet, who always encouraged me in my quest for balance between family and my engineering career, and Maribeth Bradfield, who got me started writing problems in the first place. I am also indebted to Valoree Eikinas, Robert Macia, and all the engineers at Stewart Engineering who "tried out" the problems in this book, and to Thomas H. Miller, PhD, PE, for his work in completing the technical review. I also recognize several individuals for their assistance in updating past editions. Specifically, I would like to thank Nikola Zisi for his work on the concrete problems and Fred Roland, PE, for his assistance with the steel problems.

For this third edition, I would like to thank Jennifer Tanner Eisenhauer, PhD, PE, for her assistance in updating the concrete problems and Matt Yerkey, PE, for his contributions to the new bridge problems. I would also like to thank Majid Baradar, PE, James Giancaspro, PhD, PE, and Elaine Huang, PE, for contributing seismic, structural, and construction problems, respectively, to help round out topic coverage in this edition.

On PPI's staff, my thanks go to Sarah Hubbard, director of product development and implementation; Cathy Schrott, production services manager; Julia White, editorial project manager; William Bergstrom, Lisa Devoto Farrell, Tyler Hayes, Chelsea Logan, Magnolia Molcan, and Bonnie Thomas, copy editors; Kate Hayes, production associate; Tom Bergstrom, illustrator; and Amy Schwertman La Russa, cover designer. Lastly, I thank God for giving me the talents to pursue this.

Christine A. Subasic, PE

Introduction

EXAM FORMAT

The Structural SE exam is offered in two components. The first component—vertical forces (gravity/other) and incidental lateral forces—takes place on a Friday. The second component—lateral forces (wind/earthquake)—takes place on a Saturday. Each component comprises a morning breadth and an afternoon depth module, as outlined in Table 1.

The morning breadth modules are each four hours and contain 40 multiple-choice problems that cover a range of structural engineering topics specific to vertical and lateral forces. The afternoon depth modules are also each four hours, but instead of multiple-choice problems, they contain essay problems. You may choose either the bridges or the buildings depth module, but you must work the same depth module across both exam components. That is, if you choose to work buildings for the lateral forces component, you must also work buildings for the vertical forces component.

According to NCEES, the vertical forces (gravity/other) and incidental lateral forces depth module in buildings covers loads, lateral earth pressures, analysis methods, general structural considerations (e.g., element design), structural systems integration (e.g., connections), and foundations and retaining structures. The depth module in bridges covers gravity loads, superstructures, substructures, and lateral loads other than wind and seismic. It may also require pedestrian bridge and/or vehicular bridge knowledge.

The lateral forces (wind/earthquake) depth module in buildings covers lateral forces, lateral force distribution, analysis methods, general structural considerations (e.g., element design), structural systems integration (e.g., connections), and foundations and retaining structures. The depth module in bridges covers gravity loads, superstructures, substructures, and lateral forces. It may also require pedestrian bridge and/or vehicular bridge knowledge.

For questions involving structural steel design, candidates may choose to use either the allowable strength design (ASD) method or the load and resistance factor design (LRFD) method.

For further information and tips on how to prepare for the SE exam, consult PPI's *Structural Engineering Reference Manual* or visit **www.ppi2pass.com/structural**.

THIS BOOK'S ORGANIZATION

Six-Minute Solutions for Structural Engineering (SE) Exam Morning Breadth Problems is organized into two sections: vertical forces and lateral forces. These sections correspond to the vertical and lateral forces breadth (morning) portions of the SE exam and are further divided into three subtopics: analysis of structures, design and details of structures, and construction administration. These subtopics correlate with the topics given in the NCEES SE exam specifications.

Most of the problems are quantitative, requiring calculations to arrive at a correct solution. A few are non-quantitative. Some problems will require a little more than six minutes to answer and others a little less. On average, you should expect to spend six minutes per problem.

HOW TO USE THIS BOOK

In *Six-Minute Solutions for Structural Engineering (SE) Exam Morning Breadth Problems*, each problem statement, with its supporting information and answer choices, is presented in the same format as the problems encountered on the SE breadth exam. The solutions are presented in a step-by-step sequence to help you follow the logical development of the correct solution and to provide examples of how you may want to approach your solutions as you take the SE exam.

Each problem includes a hint to provide direction in solving the problem. In addition to the correct solution, you will find an explanation of the faulty solutions leading to the three incorrect answer choices. The incorrect solutions are intended to represent common mistakes made when solving each type of problem. These may be simple mathematical errors, such as failing to square a term in an equation, or more serious errors, such as using the wrong equation.

Table 1 NCEES Structural Engineering (SE) Exam Component Module Specifications

Friday: vertical forces (gravity/other) and incidental lateral forces	

morning breadth module
    4 hours
    40 multiple-choice problems
	analysis of structures (30%)
	    loads (10%)
	    methods (20%)
	design and details of structures (65%)
	    general structural considerations (7.5%)
	    structural systems integration (2.5%)
	    structural steel (12.5%)
	    light gage/cold-formed steel (2.5%)
	    concrete (12.5%)
	    wood (10%)
	    masonry (7.5%)
	    foundations and retaining structures (10%)
	construction administration (5%)
	    procedures for mitigating nonconforming work
	    inspection methods

afternoon depth module[a]
    4 hours
    essay problems
	buildings[b]
	    steel structure (1-hour problem)
	    concrete structure (1-hour problem)
	    wood structure (1-hour problem)
	    masonry structure (1-hour problem)
	bridges
	    concrete superstructure (1-hour problem)
	    steel superstructure (2-hour problem)
	    other elements of bridges (e.g., culverts, abutments, and retaining walls) (1-hour problem)

Saturday: lateral forces (wind/earthquake)	

morning breadth module
    4 hours
    40 multiple-choice problems
	analysis of structures (37%)
	    lateral forces (10%)
	    lateral force distribution (22%)
	    methods (5%)
	design and detailing of structures (60%)
	    general structural considerations (7.5%)
	    structural systems integration (5%)
	    structural steel (10%)
	    light gage/cold-formed steel (2.5%)
	    concrete (12.5%)
	    wood (7.5%)
	    masonry (7.5%)
	    foundations and retaining structures (7.5%)
	construction administration (3%)
	    structural observation

afternoon depth module[a]
    4 hours
    essay problems
	buildings[c]
	    steel structure (1-hour problem)
	    concrete structure (1-hour problem)
	    wood and/or masonry structure (1-hour problem)
	    general analysis (e.g., existing structures, secondary structures, nonbuilding structures, and/or computer verification) (1-hour problem)
	bridges
	    columns (1-hour problem)
	    footings (1-hour problem)
	    general analysis (e.g., seismic and/or wind) (2-hour problem)

[a] Afternoon sessions focus on a single area of practice. You must choose *either* the buildings or bridges depth module, and you must work the same depth module across both exam components.
[b] At least one problem will contain a multistory building, and at least one problem will contain a foundation.
[c] At least two problems will include seismic content with a seismic design category of D or above. At least one problem will include wind content with a base wind speed of at least 110 mph. Problems may include a multistory building and/or foundation.

To optimize your study time and obtain the maximum benefit from the practice problems, consider the following suggestions.

1. Complete an overall review of the problems and identify the subjects that you are least familiar with. Work a few of these problems to assess your general understanding of the subjects and to identify your strengths and weaknesses.

2. Locate and organize relevant resource materials. (See the Codes and References section of this book for guidance.) As you work through the problems, some of these resources will emerge as more useful to you than others. These are what you will want to have on hand when taking the SE exam.

3. Work the problems in one subject area at a time, starting with the subject areas that you have the most difficulty with.

4. When possible, work problems without utilizing the hint. Always attempt your own solution before looking at the solutions provided in the book. Use the solutions to check your work or to provide guidance in finding solutions to the more difficult problems. Use the incorrect solutions to help identify pitfalls and to develop strategies to avoid them.

5. Use each subject area's solutions as a guide to understanding general problem-solving approaches. Although problems identical to those presented in *Six-Minute Solutions for Structural Engineering (SE) Exam Morning Breadth Problems* will not be encountered on the SE exam, the approach to solving problems will be the same.

Solutions presented for each example problem may represent only one of several methods for obtaining a correct answer. Although most of these problems have unique solutions, alternative problem-solving methods may produce a different, but nonetheless appropriate, answer.

Codes and References

The Structural Engineering Reference Manual is the minimum recommended library for the SE exam. The problems and solutions in this *Six-Minute Solutions* book are based on the following codes, current to the NCEES specifications at the time of publication. Since the codes used on the exam are subject to change without notice, refer to PPI's website at **www.ppi2pass.com/structural** for the most current list of exam-adopted codes.

CODES

American Association of State Highway and Transportation Officials (AASHTO). *AASHTO LRFD Bridge Design Specifications*, 5th ed. 2010.

American Concrete Institute. *Building Code Requirements for Structural Concrete* (ACI 318). 2008.

American Forest & Paper Association/American Wood Council. *National Design Specification for Wood Construction ASD/LRFD, with supplement* (NDS). 2005.

American Forest & Paper Association. *Special Design Provisions for Wind and Seismic, with Commentary* (NDS). 2008.

American Institute of Steel Construction (AISC). *Seismic Design Manual.* 2nd printing (2006) or 3rd printing (2008).

American Institute of Steel Construction (AISC). *Steel Construction Manual*, 13th ed. 2005.

American Iron and Steel Institute (AISI). *North American Specification for the Design of Cold-Formed Steel Structural Members.* 2007.

American Society of Civil Engineers. *Minimum Design Loads for Buildings and Other Structures* (ASCE7). 2005.

International Code Council. *International Building Code* (IBC). 2009.

Masonry Standards Joint Committee (*MSJC*). *Building Code Requirements for Masonry Structures* (TMS 402/ACI 530, ASCE5). 2008.

Masonry Standards Joint Committee (*MSJC*). *Specification for Masonry Structures* (TMS 602/ACI 530.1, ASCE6). 2008.

Precast/Prestressed Concrete Institute (PCI). *PCI Design Handbook: Precast and Prestressed Concrete*, 6th ed. 2004.

REFERENCES

The following references were used to prepare this book. You may also find these to be useful resources for the exam.

American Institute of Steel Construction. *Seismic Provisions for Structural Steel Buildings* (AISC 341).

American Institute of Timber Construction. *Standard Specification for Structural Glued Laminated Timber of Softwood Species* (AITC 117).

ASTM International. *Standard Specification for Hollow Brick (Hollow Masonry Units Made from Clay or Shale)* (ASTM C652). ASTM International.

Bowles, Joseph E. *Foundation Analysis and Design.* New York: McGraw-Hill.

Masonry Society, The. *Masonry Designers' Guide* (MGD). Boulder, CO: The Masonry Society.

McCormac, Jack C. *Structural Analysis.* New York: Harper & Row.

National Concrete Masonry Association (NCMA). *Concrete Masonry Wall Weights* (TEK 14-13B).

Nilson, Arthur H., David Darwin, and Charles W. Dolan. *Design of Concrete Structures.* New York: McGraw-Hill.

Occupational Safety and Health Administration. *Recording and Reporting Occupational Injuries and Illness*, 29 CFR Part 1904 (OSHA Std. 1904).

Nomenclature

a	dimension	ft	m
A	area	in^2, ft^2	mm^2, m^2
A	cross-sectional area	in^2, ft^2	mm^2, m^2
A_{cp}	area enclosed by outside perimeter of concrete	in^2	mm^2
A_t	cross-sectional area of torsion reinforcement	in^2	mm^2
b	dimension	in	mm
b	dimension from web to centerline of bolt hole in hanger connection	in	mm
b	width	in	mm
b	width of footing	ft	m
b_a	tension per bolt	kips, lbf	N
b_e	effective width of slab	in	mm
b_v	shear per bolt	kips, lbf	N
b_x, b_y	bending coefficients	$(ft\text{-}kips)^{-1}$	$(N \cdot m)^{-1}$
B	allowable tension per bolt	kips	N
B	width of column base plate in direction of column flange	in	mm
B	width of footing	ft	m
B_a	allowable tension per bolt	kips	N
B_v	allowable shear per bolt	lbf	N
c	neutral axis depth	in	mm
c	undrained shear strength (cohesion)	lbf/ft^2	Pa
c_A	adhesion	lbf/ft^2	Pa
C	correction factor	various	various
C_1	moment coefficient	–	–
C_D	drag coefficient	–	–
d	beam depth	in	mm
d	beam depth (to tension steel centroid)	in	mm
d	bolt diameter	in	mm
d	diameter	in, ft	mm, m
d	effective depth	in	mm
d'	diameter of bolt hole	in	mm
d_b	nominal diameter of reinforcing bar	in	mm
D	dead load	kips, lbf, lbf/ft^2	N, Pa
D	depth	ft, in	m
D	diameter	ft	m
DF	distribution factor	–	–
e	eccentricity	in, ft	mm, m
E	earthquake load	kips	N
E	modulus of elasticity	$kips/in^2$, lbf/ft^2, lbf/in^2	Pa
E'	allowable modulus of elasticity	$kips/in^2$, lbf/ft^2, lbf/in^2	Pa
f	stress	lbf/in^2	Pa
f_a	compressive stress in masonry due to axial load alone	lbf/in^2	Pa
f_b	bending stress	lbf/in^2	Pa
f_b	stress in masonry due to flexure alone	lbf/in^2	Pa
f'_c	compressive strength of concrete	lbf/in^2	Pa
f'_m	compressive strength of masonry	lbf/in^2	Pa
f_r	modulus of rupture	lbf/in^2	Pa
f_s	shear stress	lbf/in^2	Pa
f_s	stress in steel reinforcement	lbf/in^2	Pa
f_t	tensile stress	lbf/in^2	Pa
f_v	shear stress in masonry	lbf/in^2	Pa
f_y	yield stress	lbf/in^2	Pa
F	allowable stress	$kips/in^2$	Pa
F	factor of safety	–	–
F	force	lbf, kips	N
F	strength	$kips/in^2$	Pa
F'	reduced allowable stress	$kips/in^2$	Pa
F_a	allowable compressive stress due to axial load alone	lbf/in^2	Pa
F_b	allowable bending stress	lbf/in^2	Pa
F_b	allowable compressive stress due to flexure alone	lbf/in^2	Pa
F_p	allowable bearing pressure	lbf/in^2	Pa
F_s	allowable tensile stress in steel reinforcement	lbf/in^2	Pa
F_t	allowable tensile stress	lbf/in^2	Pa
F_{TH}	threshold stress range	$kips/in^2$	Pa
F_v	allowable shear stress in masonry	lbf/in^2	Pa
F_y	yield strength of steel	lbf/in^2	Pa
FEM	fixed-end moment	ft-kips	$N \cdot m$
g	lateral gage spacing of adjacent holes	in	mm
g	ratio of distance between tension steel and compression steel to overall column depth	–	–
G	gust factor	–	–
h	distance	ft	m
h	effective height	in	mm
h	height	in, ft	mm, m
h	overall thickness	in, ft	mm, m

Symbol	Description	US Units	SI Units
h'	modified overall thickness	in, ft	mm, m
H	distance from soil surface to footing base	ft	m
H_s	length of stud connector	in	mm
I	moment of inertia	in^4	mm^4
I_p	polar moment of inertia	in^4	mm^4
j	ratio of distance between centroid of flexural compressive forces and centroid of tensile forces to depth	–	–
k	coefficient	–	–
k	coefficient of lateral earth pressure	–	–
k	effective length factor	–	–
k	stiffness	lbf/ft	N/m
K	coefficient	–	–
K	effective length factor	–	–
K	relative stiffness	ft^{-1}	m^{-1}
l	distance between points of lateral support of compression member in a given plane	in, ft	mm, m
l	length	in, ft	mm, m
l	span length	in, ft	mm, m
l_b	effective embedment length of headed or bent anchor bolts	in	mm
l_c	vertical distance between supports	in	mm
l_d	development length of reinforcement	in	mm
l_h	distance from center of bolt hole to beam end	in	mm
l_n	clear span length	in, ft	mm, m
l_u	unsupported length of a compression member	in	m
l_v	distance from center of bolt hole to edge of web	in	mm
L	length	in, ft	m
L	length of footing	ft	m
L	live load	kips, lbf, lbf/ft^2	N, Pa
L	span length	in, ft	mm, m
M	moment	ft-kips, in-lbf, ft-lbf	N·mm, N·m
M_a	maximum moment	in-lbf	N·m
M_m	resisting moment assuming masonry governs	in-lbf	N·m
M_o	total factored static moment	in-lbf	N·m
M_R	beam resisting moment	ft-kips	N·mm
M_s	resisting moment assuming steel governs	in-lbf	N·m
M_t	torsional moment	ft-kips	N·mm
n	modular ratio	–	–
n	quantity (number of)	–	–
N	bearing capacity factor	–	–
N	length of column base plate in direction of column depth	in	mm
N_r	number of studs in one rib	–	–
p	perimeter	in, ft	mm, m
p	pressure	lbf/ft^2	Pa
p_{cp}	outside perimeter of concrete	in	mm
P	axial load	kips	N
P	load	kips	N
P	prestress force in a tendon	lbf, kips	N
P_a	allowable compressive force in reinforced masonry due to axial load	kips	N
P_e	Euler buckling load	kips	N
P_{nw}	nominal axial strength of a wall	kips	N
P_o	maximum allowable axial load for zero eccentricity	kips	N
q	allowable horizontal shear per shear connector	kips	N
q	soil pressure under footing	lbf/ft^2	Pa
q	uniform surcharge	lbf/ft, lbf/ft^2	N/m, Pa
q_c	tip resistance	lbf/ft^2	Pa
q_s	skin friction resistance	lbf/ft^2	Pa
Q	bearing capacity	lbf/ft^2	Pa
Q	nominal load effect	various	various
Q	statical moment	in^3	mm^3
r	radius of gyration	in, ft	mm, m
r	rigidity	–	–
R	allowable load per bolt	kips	N
R	concentrated load	lbf, kips	N
R	reaction	lbf, kips	N
R	resultant force	lbf	N
RF	reduction factor	–	–
s	spacing	in	mm
S	force	lbf	N
S	section modulus	in^3, ft^3	mm^3, m^3
S	snow load	kips	N
S_{DS}	design earthquake spectral response accelerations at short period	–	–
S_{MS}	maximum considered earthquake spectral response accelerations at short period	–	–
t	nominal weld size	in	mm
t	slab thickness	in	mm
t	thickness	in, ft	mm, m
t	wall thickness	in, ft	mm, m
t_e	effective throat thickness of a weld	in	mm
T	period	sec	s
T	temperature	°F, °R	°C, K
T	tension force	kips, lbf	N
T	torsional moment (torque)	ft-lbf	N·m
T_u	factored torsional moment	ft-lbf	N·m
u	unit force	–	–
U	ultimate strength required to resist factored loads	lbf	N
v	shear stress	lbf/in^2	Pa
v	wind speed	mph	kph
v_c	allowable concrete shear stress	lbf/in^2	Pa

Symbol	Description	US units	SI units
v_u	shear stress due to factored loads	lbf/in²	Pa
V	design shear force	lbf	N
V	shear	kips, lbf	N
V	shear strength	lbf, lbf/in²	N, Pa
V_c	allowable concrete shear strength	lbf	N
V_u	factored shear force	lbf	N
w	distributed load	lbf/ft	N/m
w	tributary width	in, ft	mm, m
w	uniformly distributed load	lbf/ft, lbf/ft²	N/m, Pa
w	weld size	in	mm
w'	tributary width	in, ft	mm, m
W	effective seismic weight	lbf, kips	N
W	nail withdrawal value	lbf	N
W	weight	lbf	N
W	wind load	lbf	N
x	distance	in, ft	mm, m
x	location	in	mm
x	x-coordinate of position	ft	m
$\bar{x}$	distance to center of rigidity in the x-direction	in, ft	mm, m
$\bar{y}$	distance from centroidal axis to the centroid of the area	in	mm
$\bar{y}$	distance to center of rigidity in the y-direction	in, ft	mm, m
y	location	in	m
y	y-coordinate of position	ft	m
y_c	distance from top of the section to the centroid of the section	in	mm
y_t	distance from centroid of the section to the extreme fiber in tension	in	mm
Z	connecter lateral design value	lbf	N

Symbols

Symbol	Description	US units	SI units
α	coefficient of linear thermal expansion	°F⁻¹	°C⁻¹
α	ratio of flexural stiffness of beams in comparison to slab	–	–
β	column strength factor	–	–
β	ratio of clear spans in long- to-short directions of a two-way slab	–	–
β	ratio of long side to short side of a footing	–	–
γ	ratio of the distance between bars on opposite faces of a column to the overall column dimension, both measured in the direction of bending	–	–
γ	specific weight (unit weight)	lbf/ft³	N/m³
Δ	deflection	in	mm
θ	angle	deg, rad	deg, rad
λ	distance from centroid of compressed area to extreme compression fiber	in	mm
λ	height and exposure adjustment factor	–	–
ρ	reinforcement ratio	–	–
ρ_g	longitudinal reinforcement ratio	–	–
ρ_h	ratio of horizontal shear reinforcement to gross concrete area	–	–
ρ_n	ratio of vertical shear reinforcement area to gross concrete area for a shear wall	–	–
σ	normal stress	lbf/ft²	Pa
τ	shear stress	lbf/ft²	Pa
ϕ	strength reduction factor	–	–
Ψ	relative stiffness parameter	–	–

Subscripts

γ	density
0	initial
a	active, allowable, or due to axial loading
A	adhesion
b	beam, bending, bolt, or masonry breakout
bm	beam
bot	bottom
BS	block shear
c	cohesive, column, concrete, curvature, or masonry crushing
cr	cracked or cracking
cs	critical section
d	directionality or penetration depth
D	dead load or duration
e	effective or exposure
eg	end grain
E	Euler
f	flange, flat, force, form, or skin friction
fs	face shell
fu	flat use
F	size
g	gross or grout
h	horizontal
i	initial, inside, or ith member
j	jth member
k	kern or kth member
l	longitudinal
L	beam stability, left, or live load
m	masonry
max	maximum
min	minimum
M	wet service
n	nail, net, or nominal
ns	nonsway frame
o	centroidal or initial
OT	overturning
p	bearing, passive, pile tip, plate, or prestressed
prov	provided
q	surcharge

r	rafter, reduced, repetitive, resultant, rib, roof, or rupture	th	thermal
		tr	transformed
R	resistance, resisting, resistive, resultant, or right	u	ultimate (factored), ultimate tensile, unbraced, or untreated
req	required		
s	shear, side friction, simplified, skin, snow, spiral, steel reinforcement, or steel yielding	v	shear or vertical
		V	volume
sat	saturated	w	wall, web, weld, or wind
sd	short direction	x	at a distance x, in x-direction, or strong axis
sp	single pile	y	in y-direction, weak axis, or yield
sr	stress range	z	at height z
SL	sliding		
t	temperature, tensile, tension, topography, torsion, or tributary		

Vertical Breadth Problems

Analysis of Structures

1. A single-story flex warehouse building is constructed of 8 in concrete masonry walls that have been grouted solid. One 20 ft long wall contains a 4 ft opening centered in the wall as shown. A steel lintel made from two 5 in × $3^1/_2$ in × $^5/_{16}$ in angles spans the opening. A 700 lbf concentrated load from a roof truss is centered over the opening. The unit weight of the wall is 150 lbf/ft^3.

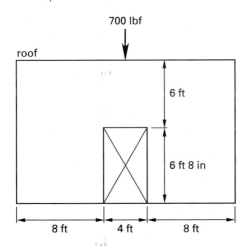

Using ASD, what is most nearly the design moment on the lintel?

(A) 300 ft-lbf

(B) 400 ft-lbf

(C) 1000 ft-lbf

(D) 1200 ft-lbf

Hint: The wall described will exhibit arching action over the opening.

2. A steel beam supports a 6 in lightweight concrete masonry (CMU) wall around a mechanical room. The CMU has a unit weight, γ, of 85 lbf/ft^3. If the wall is 8 ft high and 10 ft long, the load on the beam is most nearly

(A) 340 lbf/ft

(B) 680 lbf/ft

(C) 3400 lbf/ft

(D) 4100 lbf/ft

Hint: The load on the beam is uniformly distributed.

3. A two-story office building located in a South Dakota office park is 40 ft × 60 ft in plan. The asphalt-shingle hip roof has a 6:12 pitch and is supported by rafters spanning the short direction of the building. The attic is vented and insulated with R-30 insulation. The local building code official requires design using the IBC. If the ground snow load is 40 lbf/ft^2, what is most nearly the maximum leeward snow load?

(A) 31 lbf/ft^2

(B) 32 lbf/ft^2

(C) 40 lbf/ft^2

(D) 46 lbf/ft^2

Hint: The leeward side of the roof is the side away from the wind.

4. A 2.5 in diameter mild steel bar is fixed at each end by a steel plate. The modulus of elasticity of the steel is 29×10^6 lbf/in^2. At 8:00 a.m., the bar has a temperature of 35°F and experiences zero stress. At 3:30 p.m., the temperature of the bar is 100°F. Ignore gravity loads. The axial load in the bar at 3:30 p.m. is most nearly

(A) 60 kips

(B) 92 kips

(C) 240 kips

(D) 6000 kips

Hint: Chapter 2 of the AISC *Steel Construction Manual* contains information on section properties and thermal expansion properties of steel.

5. The plain concrete foundation wall shown supports a 12 in nominal concrete masonry unit (CMU) wall. The foundation wall bears on soil composed of poorly graded clean sands and is laterally supported top and bottom.

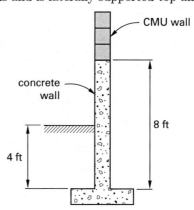

Using the prescriptive criteria found in the IBC, the minimum thickness of the foundation wall is most nearly

(A) 7.5 in

(B) 8.0 in

(C) 9.5 in

(D) 12 in

Hint: Use IBC Sec. 1805 to determine the minimum thickness.

6. A flat-plate concrete slab is supported by 12 in square concrete columns. The floor dead load is 100 lbf/ft^2. The floor live load is 30 lbf/ft^2. The design moments of the interior slab panel shown are to be determined using the direct design method.

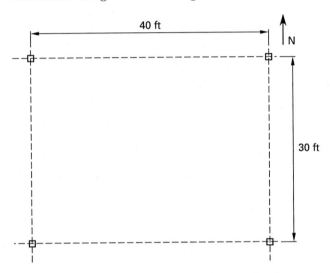

The midspan moment of the middle strip in the east/west direction is most nearly

(A) 77 ft-kips

(B) 99 ft-kips

(C) 130 ft-kips

(D) 140 ft-kips

Hint: Refer to ACI 318 Chap. 13.

7. Analyze the truss shown.

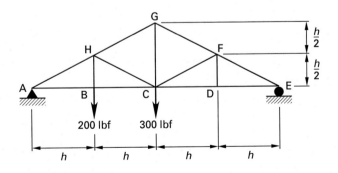

The force in member AH is most nearly

(A) 300 lbf (compression)

(B) 600 lbf (compression)

(C) 670 lbf (compression)

(D) 670 lbf (tension)

Hint: Solve by using the method of joints.

8. A pin-connected truss is used to support a sign as shown. The product of the area and the modulus of elasticity equals 3 kips for all members except AD and BC, for which the product of the area and the modulus of elasticity equals 5 kips.

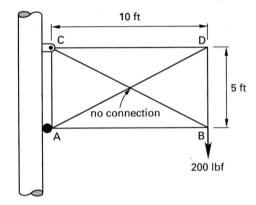

What is most nearly the force in member BC?

(A) −210 lbf (compression)

(B) 230 lbf (tension)

(C) 240 lbf (tension)

(D) 450 lbf (tension)

Hint: Assess whether the truss is determinate.

9. The T-shaped beam shown carries a 100 lbf shear load.

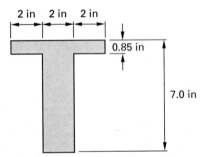

The shear stress at the centroid of the cross section is most nearly

(A) 6.5 lbf/in^2

(B) 8.1 lbf/in^2

(C) 10 lbf/in^2

(D) 22 lbf/in^2

Hint: First, find the centroid of the cross section.

10. A beam with a constant cross section and constant modulus of elasticity is loaded as shown. The relative stiffness of section AB is 0.286/ft and of section BC is 0.190/ft.

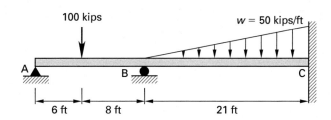

Using moment distribution, the moment at the fixed end is most nearly

(A) 1100 ft-kips, counterclockwise

(B) 1200 ft-kips, clockwise

(C) 1200 ft-kips, counterclockwise

(D) 1300 ft-kips, clockwise

Hint: Consult a reference containing fixed-end moments.

11. A moment distribution analysis of a beam determined the moments at the beam supports as shown.

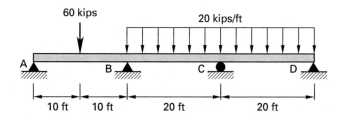

	A	B		C		D
DF	1.0	0.5	0.5	0.5	0.5	1.0
FEM	−150.0	+150.0	−666.7	+666.7	−666.7	+666.7

Moment distribution gives the following results.

	A	B		C		D
M	0	+386.8	−386.8	+903.2	−903.2	0

The reaction at support B is most nearly

(A) 50 kips

(B) 120 kips

(C) 170 kips

(D) 220 kips

Hint: Use free-body diagrams to determine the shear at the supports.

12. The beam shown has a modulus of elasticity of 1.2×10^6 lbf/in^2 and a moment of inertia of 240 in^4.

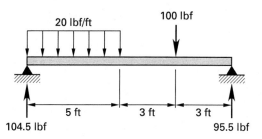

Using the conjugate beam method, the deflection at a point 5 ft from the left support of the beam is most nearly

(A) 1.1×10^{-5} in

(B) 1.5×10^{-4} in

(C) 2.2×10^{-2} in

(D) 3.1×10^{-2} in

Hint: Begin by drawing the M/EI diagram for the beam.

13. A 20 ft simply supported beam carries a load that decreases linearly from a maximum of 350 lbf/ft at its left support to 0 at the midspan of the beam. Measured from the left support, the point at which the shear is zero is most nearly

(A) 0.87 ft

(B) 4.1 ft

(C) 5.9 ft

(D) 14 ft

Hint: Begin by drawing the free-body diagram.

14. A 12 in unreinforced concrete masonry wall supports joists spaced 12 in on center. The reaction from each joist is 700 lbf. The wall is grouted solid.

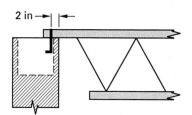

The stress on the wall due to the joists is most nearly

(A) 5.0 lbf/in^2 (compression)

(B) 5.0 lbf/in^2 (tension)

(C) 5.0 lbf/in^2 (compression), 10 lbf/in^2 (tension)

(D) 15 lbf/in^2 (compression), 5.0 lbf/in^2 (tension)

Hint: Determine the eccentricity of the load.

Design and Detailing of Structures

15. A concrete slab designed for a parking garage constructed in Chicago uses normal-weight concrete with a compressive strength of 3000 lbf/in^2. If the maximum aggregate size is 1 in, the total percentage of air content of the concrete mix should be

(A) 0.06%

(B) 0.45%

(C) 4.5%

(D) 6.0%

Hint: The parking garage in Chicago is exposed to freezing and thawing conditions.

16. A reinforced concrete building has a 6 in flat-plate slab on each floor and 12 in diameter columns. The floor-to-floor distance is 13 ft. A column has a factored axial load of 300 kips and equal factored end moments of 100 ft-kips and is subjected to double curvature. The modulus of elasticity of concrete is 3.6×10^6 lbf/in^2. If the column does not have any transverse loads and is not subject to sway, the design moment is most nearly

(A) 100 ft-kips

(B) 4000 ft-kips

(C) 9000 ft-kips

(D) 20,000 ft-kips

Hint: Use magnified moments to determine the design moment.

17. A two-story office building has a 60 ft × 100 ft rectangular floor plan. Steel columns spaced 20 ft apart carry the following loads from the roof and second floor. Live load reductions are not permitted.

roof dead load	15 lbf/ft^2
roof live load	20 lbf/ft^2
floor dead load	15 lbf/ft^2
floor live load	60 lbf/ft^2

What is most nearly the total service load on an interior first-floor column?

(A) 1.8 kips

(B) 14 kips

(C) 36 kips

(D) 44 kips

Hint: Determine the tributary area for an interior column.

18. A single-story steel-framed building has columns spaced 15 ft on center in the north/south direction and 20 ft on center in the east/west direction. The columns support a uniform dead load of 40 lbf/ft^2. The building has an ordinary roof with a 1:2 pitch. Using the IBC, the minimum live load on an interior column is most nearly

(A) 4.1 kips

(B) 4.9 kips

(C) 6.0 kips

(D) 17 kips

Hint: IBC Sec. 1607 covers live loads.

19. The cantilevered beam shown has a varying moment of inertia. The moment of inertia for the first 15 ft, I_1, is 2000 in^4. The second moment of inertia, I_2, is 1000 in^4. The modulus of elasticity of the beam is 29×10^6 lbf/in^2.

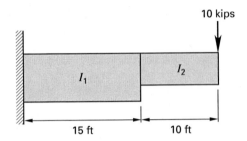

The deflection at the free end of the beam is most nearly

(A) 0.0017 in

(B) 1.7 in

(C) 3.0 in

(D) 3.1 in

Hint: Use the moment-area method to find the deflection of the beam.

20. A church is built with a wood roof that is exposed on the interior. The roof beam is 60 ft in length. According to the IBC, what is the total deflection limit for the roof beam?

(A) 0.5 in

(B) 3 in

(C) 4 in

(D) 6 in

Hint: Refer to IBC Sec. 1604.3.

21. A steel channel strut is connected with high-strength bolts with end-bearing connections. The strut is subjected to a service dead load tensile force of 20 kips. The service live load varies from a 10 kip compressive force to a 50 kip tensile force. It is estimated that the live load may be applied 200 times per day for the life of the structure. The structure is expected to last at least 25 years. The structure is subject to normal atmospheric conditions with temperatures less than 300°F. If the channel is ASTM A36 steel, what is the lightest section that can carry the load?

(A) C7 × 9.8

(B) C6 × 13

(C) C8 × 11.5

(D) C8 × 18.75

Hint: Reversal of the live load must be considered in the design.

22. A simply supported composite ASTM A992 steel beam has the following properties.

area of concrete	288 in^2
weight of concrete	110 lbf/ft^3
area of steel	11.8 in^2
compressive strength of concrete	3000 lbf/in^2
yield stress of steel	50 kips/in^2

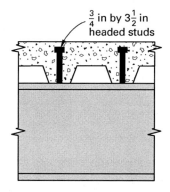

The composite deck has a nominal rib height of 2 in, an average rib width of 3 in, and one 3/4 in diameter stud per rib in the strong position. The total number of studs needed for full composite action is most nearly

(A) 35 studs

(B) 58 studs

(C) 70 studs

(D) 86 studs

Hint: Use Chap. I of the AISC *Steel Construction Manual*.

23. A steel column supports an unfactored concentric dead load of 130 kips and an unfactored concentric live load of 390 kips. The effective length with respect to the major axis is 32 ft. The effective length with respect to the minor axis is 18 ft. Using LRFD, what is the lightest ASTM A992 W shape that can be used for the column if the column depth cannot exceed 12 in (nominal)?

(A) W12 × 87

(B) W12 × 136

(C) W12 × 170

(D) W12 × 252

Hint: Use the columns tables in Part 4 (Column Design) of the AISC *Steel Construction Manual*.

24. The base plate of a W12 × 72 ASTM A992 structural steel column bears directly on an 8 ft × 8 ft concrete spread-footing. The base plate is made of ASTM A36 steel and is limited in size to 14 in × 16 in. The column supports a dead load of 105 kips and a live load of 315 kips.

compressive strength of concrete	3000 lbf/in^2
yield stress of steel	36 kips/in^2

Using LRFD and the AISC *Steel Construction Manual*, the thickness of the base plate is most nearly

(A) 0.916 in

(B) 0.997 in

(C) 1.27 in

(D) 1.34 in

Hint: Start by determining the bearing stress on the concrete.

25. A W14 × 109 steel column carries an axial load of 320 kips and a moment of 200 ft-kips about the y-axis. The centerline of the anchor bolts is 1.5 in outside the column flanges as shown.

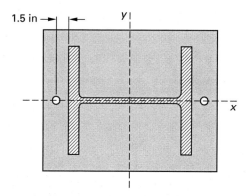

If A307 anchor bolts are used, what size bolt is needed using ASD?

(A) 5/8 in diameter

(B) 3/4 in diameter

(C) 7/8 in diameter

(D) 1 in diameter

Hint: Columns with relatively large moments may be subject to uplift on the base plate.

26. The three-story, 4 in brick veneer (40 lbf/ft²) office building shown is framed with ASTM A992 structural steel. Assume simple connections and full lateral support of the beam. The floor dead load is 60 lbf/ft², and the floor live load is 40 lbf/ft². Total deflection should not exceed $L/600$ or 0.3 in. Design using the AISC ASD method.

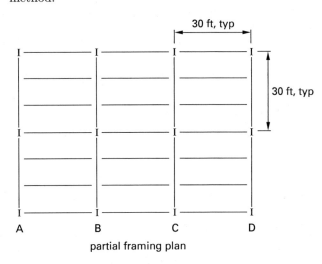

partial framing plan

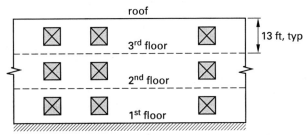

partial facade elevation

What is the minimum size of third-floor spandrel beam BC for a typical 30 ft × 30 ft bay?

(A) W12 × 19

(B) W14 × 30

(C) W21 × 55

(D) W24 × 84

Hint: The weight of the facade is typically supported by spandrel beams at each floor.

27. A beam-column made from ASTM A992 steel supports an axial load of 400 kips at an eccentricity of 12 in about the strong axis. The effective length, KL, is 25 ft. Using ASD, select a W shape to carry the load.

(A) W14 × 120

(B) W14 × 159

(C) W14 × 193

(D) W14 × 311

Hint: Use the equivalent axial load procedure found in Part 4 and Part 6 of the AISC *Steel Construction Manual*.

28. An HSS $4 \times 4 \times {}^3\!/_8$ tube beam is welded to the flange of a W10 × 33 ASTM A36 steel column as shown. The reaction at the column is 12,000 lbf. The welds are fillet welds made using the shielded metal arc welding (SMAW) process with E70XX electrodes.

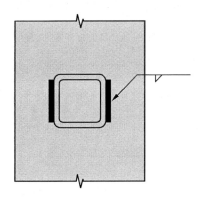

What is the required size of the weld?

(A) $^1\!/_8$ in

(B) $^3\!/_{16}$ in

(C) $^1\!/_4$ in

(D) The shear exceeds the capacity of a fillet weld.

Hint: Use Sec. J2 of the AISC *Steel Construction Manual*.

29. A W14 × 22 ASTM A992 steel beam is bolted to a column flange with L3$^1\!/_2$ × 3$^1\!/_2$ × $^5\!/_{16}$ double angles as shown. A single row of $^3\!/_4$ in diameter ASTM A325-N bolts is used. Standard size holes are used. Assume that the column-to-clip angle connection is satisfactory.

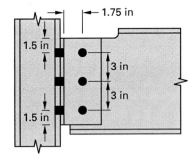

(not to scale)

Using LRFD and the AISC *Steel Construction Manual*, the maximum beam reaction is most nearly

(A) 32 kips

(B) 45 kips

(C) 48 kips

(D) 95 kips

Hint: Refer to Part 10 of the AISC *Steel Construction Manual* for connection design.

30. An L4 × 8 × 1/2 angle (LLV) is welded to a W12 × 50 column using the shielded metal arc welding process and E70XX electrodes.

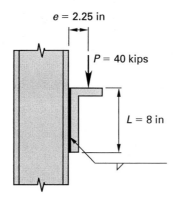

What is the required fillet weld size if the angle is welded on both sides of the vertical leg only?

(A) 1/8 in
(B) 3/16 in
(C) 1/4 in
(D) 5/16 in

Hint: The weld is subject to both shear and bending.

31. A truss made from ASTM A36 steel is used to carry vehicular traffic. The diagonal compression members are made from two L8 × 6 × 1/2 angles with a 3/8 in gusset plate between them as shown. The long legs of the angle are back-to-back. The controlling slenderness ratio, Kl/r, is 110, and the effective length, Kl_x, is 23 ft.

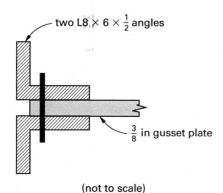

If the axial load is concentric, what is most nearly the critical compressive stress?

(A) 11 kips/in²
(B) 19 lbf/in²
(C) 19 kips/in²
(D) 20 kips/in²

Hint: The legs of the angle are unstiffened elements.

32. An ASTM A992 W18 × 40 composite beam with a 4 in concrete slab supports a live load moment of 140 ft-kips and a dead load moment of 80 ft-kips. The dead load includes the weight of the slab and beam. The construction is unshored.

section modulus of the beam	68.4 in³
transformed section modulus, measured to the bottom of the section	103 in³
transformed section modulus, measured to the top of the concrete	350 in³
transformed section modulus, measured to the steel/concrete interface	1680 in³

Using ASD, the bending stress in the bottom fibers of the steel beam due to dead load is most nearly

(A) 9.3 kips/in²
(B) 14 kips/in²
(C) 39 kips/in²
(D) 1200 lbf/in²

Hint: In unshored construction, the beam carries the full dead load.

33. Lightweight concrete (100 lbf/ft³) is used for the one-way slab shown.

compressive strength of concrete	3000 lbf/in²
yield stress of reinforcement steel	60,000 lbf/in²

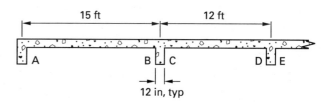

What is the minimum thickness for slab AB?

(A) 7.0 in
(B) 8.1 in
(C) 8.6 in
(D) 10 in

Hint: Minimum thicknesses of one-way slabs not supporting partitions or other elements likely to be damaged by deflection can be determined using ACI 318 Table 9.5(a).

34. A 6 in normal-weight concrete flat slab supports a factored load of 320 lbf/ft². The beam shown is a continuous two-span beam between columns.

specific weight of slab	150 lbf/ft³
α_{f1}	1.5
positive moment in the column strip	90 ft-kips
negative moment in the column strip	−220 ft-kips

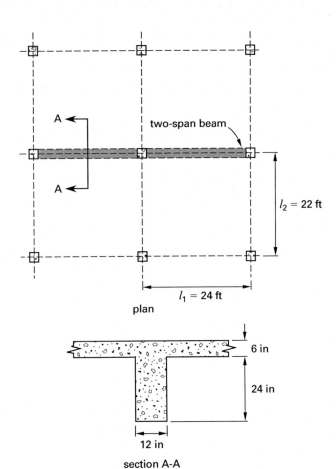

plan

section A-A

What is the maximum midspan beam moment?

(A) 77 ft-kips
(B) 89 ft-kips
(C) 91 ft-kips
(D) 290 ft-kips

Hint: Refer to ACI 318 Chap. 13.

35. A circular spiral concrete column supports a 300 kip dead load and a 350 kip live load. The concrete compressive strength is 4000 lbf/in², and the yield stress of the reinforcement is 60,000 lbf/in². If the maximum reinforcement is used, the cross-sectional area of the column is most nearly

(A) 140 in²
(B) 180 in²
(C) 200 in²
(D) 210 in²

Hint: Refer to ACI 318 Sec. 10.3.6.

36. A single-story concrete bearing wall supported at the roof and foundation has the following properties.

wall thickness	12 in
concrete compressive strength	4000 lbf/in²
effective length factor	1.0
length of wall	16 ft
uniform factored axial load	5 kips/ft

If the wall is empirically designed, its maximum height is most nearly

(A) 25.0 ft
(B) 26.9 ft
(C) 31.6 ft
(D) 32.0 ft

Hint: Refer to ACI 318 Chap. 14.

37. A three-story reinforced concrete building is supported on columns placed on a grid and spaced 20 ft apart in each direction. The story height, measured from the top of one slab to the top of the slab above, is 13 ft. At the first level, the distance from the tops of the footings to the top of the first-story slab is 18 ft. The columns are 18 in × 18 in. Beams span between columns in each direction. The slabs are all 6 in thick two-way spans. 6000 lbf/in² concrete with a modulus of elasticity of 4.4×10^6 lbf/in² is used throughout the structure.

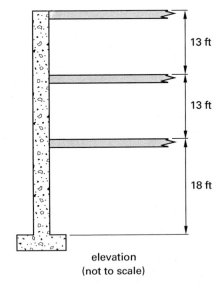

elevation
(not to scale)

If the moment of inertia of the spandrel beams is 10,000 in⁴, what is the relative stiffness parameter, Ψ, at the top of the first-floor corner column?

(A) 2.3
(B) 4.6
(C) 5.4
(D) 6.6

Hint: Refer to ACI 318 Chap. 10.

38. A two-way flat-slab system is supported by 12 in concrete columns as shown.

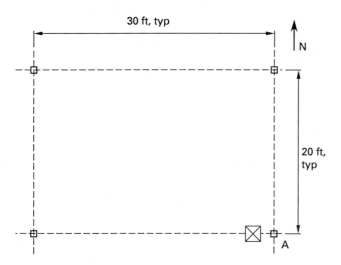

If no special analysis is used, the maximum size opening that can be located adjacent to column A, centered on the east/west column centerline, is most nearly

(A) 0.0 ft^2

(B) 0.39 ft^2

(C) 1.6 ft^2

(D) 2.3 ft^2

Hint: Refer to ACI 318 Chap. 13 on two-way slabs.

39. A prestressed concrete beam has a specified compressive strength of concrete of 6000 lbf/in^2 and the following properties.

compressive strength of concrete at time of initial prestress	4000 lbf/in^2
specified yield strength of nonprestressed reinforcement	60,000 lbf/in^2
initial prestress force	150 kips

The extreme fiber stress in tension in the concrete immediately after prestress transfer is limited to

(A) 150 lbf/in^2

(B) 190 lbf/in^2

(C) 230 lbf/in^2

(D) 380 lbf/in^2

Hint: Refer to ACI 318 Chap. 18.

40. The maximum factored shear in the 6 in × 9 in concrete beam shown is 1800 lbf. The compressive strength of concrete is 3000 lbf/in^2, and the yield stress of the steel reinforcement is 60,000 lbf/in^2.

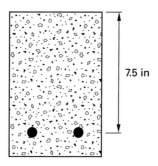

What is most nearly the required shear reinforcement?

(A) no. 3 U-stirrups at 4.0 in spacing

(B) no. 3 U-stirrups at 4.5 in spacing

(C) no. 3 U-stirrups at 24 in spacing

(D) none

Hint: Refer to ACI 318 Sec. 11.4.

41. A 35 in × 40 in reinforced concrete beam has longitudinal and torsional reinforcement with a yield stress of 40,000 lbf/in^2. The compressive strength of the concrete is 4000 lbf/in^2. The nominal shear strength of the concrete is 168.2 kips. The area enclosed by the outside perimeter of concrete is 1400 in^2. The outside perimeter of concrete cross section is 150 in.

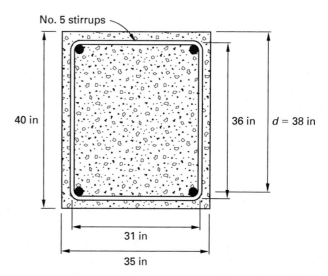

The torsion and shear diagrams are as shown. Torsion must be considered.

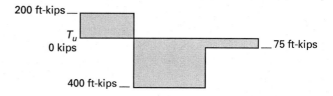

What is the maximum spacing of the no. 5 stirrups shown at the point of maximum torsion?

(A) 2.0 in
(B) 2.7 in
(C) 3.1 in
(D) 3.7 in

Hint: Refer to ACI 318 Sec. 11.6 for beams with torsion.

42. A 20 in × 30 in reinforced concrete beam has five continuous 20 ft spans. The point of inflection for positive moment is 3 ft from the right face of the support. According to ACI 318, the minimum clear cover on the reinforcing bars is 1.5 in, and the minimum clear spacing of the reinforcing bars is 2.5 in. Appropriate shear reinforcement is provided.

concrete compressive strength	6000 lbf/in²
yield stress of reinforcement	60,000 lbf/in²

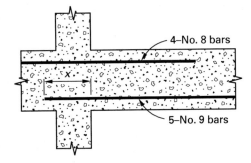

The required minimum length of the bottom bars beyond the face of the support, x, is most nearly

(A) 0 in
(B) 6 in
(C) 30 in
(D) 40 in

Hint: Refer to ACI 318 Chap. 12.

43. Using the *National Design Specification for Wood Construction* (NDS), which of the following statements is true?

I. The temperature factor, C_t, applies to members subjected to extremely cold temperatures.

II. The volume factor, C_V, applies only to glued laminated timber and structural composite lumber bending members.

III. The bending design allowable stress, F_b, for a floor framed with 1 × 6 sawn lumber joists must be multiplied by the repetitive member factor.

IV. The load duration factor, C_D, does not apply to the modulus of elasticity values.

(A) I and II
(B) II and III
(C) II and IV
(D) III and IV

Hint: Refer to NDS Sec. 2.3 for adjustment of design values.

44. Wood formwork consisting of 3 × 4 joists and stringers is used to support a 4 in normal-weight concrete slab. The live load during construction is 50 lbf/ft². Given that the maximum spacing of the stringers limited by bending in the joists is $(4466 \text{ lbf}/w)^{1/2}$, the maximum spacing of the stringers as limited by bending in the stringers is $523.4 \text{ lbf/ft}/w$, and the maximum spacing of the stringers as limited by the load to the post is $512.8 \text{ lbf/ft}/w$. The maximum spacing of the stringers is most nearly

(A) 5.1 ft
(B) 5.9 ft
(C) 6.7 ft
(D) 9.5 ft

Hint: Determine the load on the stringers.

45. A glued laminated (glulam) beam with the designation 20F-V7 is used to span a 20 ft opening. Which of the following statements is true?

(A) The beam is mechanically graded.
(B) The depth of the beam is 20 in.
(C) The shear capacity of the beam is 700 lbf/in².
(D) The flexural tensile capacity of the beam is 2000 lbf/in².

Hint: The answer does not depend upon the type of wood.

46. A built-up column made from three southern pine sawn 2 × 6 members nailed together meets the requirements of NDS Sec. 15.3.3. The ends of the column are free to rotate but are not free to translate.

distance between points of lateral support of compression member in plane 1	9.0 ft
distance between points of lateral support of compression member in plane 2	4.5 ft
E'_{min}	620,000 lbf/in²
F^*_c	1750 lbf/in²

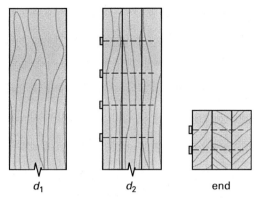

What is the critical column stability factor for this column?

(A) 0.44
(B) 0.52
(C) 0.57
(D) 0.59

Hint: NDS Sec. 15.3 covers the design of built-up columns.

47. Plywood sheathing is attached to 2×12 roof rafters with 6d box nails. The nails penetrate 1 in into the rafters. The rafters are spaced 16 in on center. All members are southern pine. Sustained temperatures do not exceed 100°F. If the uplift on the roof is 36 lbf/ft^2, the maximum nail spacing is most nearly

(A) 7.8 in
(B) 11 in
(C) 12 in
(D) 14 in

Hint: Refer to NDS Chap. 11.

48. A deck is built on the back of a house in a humid climate using the design requirements contained in the NDS. A 2×8 beam is screwed to the face of a 4×4 post as shown. The wood is southern pine. The screws shall have a penetration greater than or equal to ten times the shank diameter. The dead-load plus live-load end reaction of the beam is 605 lbf.

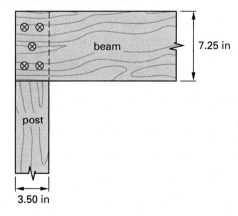

If five 12-gage wood screws are used, what is most nearly the allowable load on the connection?

(A) 450 lbf
(B) 500 lbf
(C) 560 lbf
(D) 800 lbf

Hint: Refer to NDS Chap. 11 for wood screw design values.

49. A 7.5 in wide $\times$ 16 in deep reinforced concrete masonry lintel is used to carry uniform dead and live loads of 450 lbf/ft and 550 lbf/ft, respectively. The self-weight of the lintel is 120 lbf/ft. The load due to the weight of the wall above the lintel is as shown for the lintel span length, L. The specified compressive strength of the masonry is 3000 lbf/in^2. Design the lintel using allowable stress design.

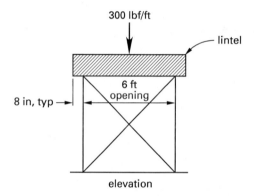

elevation

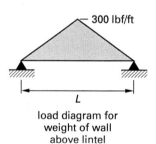

load diagram for weight of wall above lintel

If two no. 4, grade 60 reinforcing bars are used side-by-side at the bottom of the beam, what is most nearly the tensile stress in the steel?

(A) 14,000 lbf/in^2
(B) 15,000 lbf/in^2
(C) 16,000 lbf/in^2
(D) 17,000 lbf/in^2

Hint: Calculate the lintel span length first.

50. The 6 in nominal (5⅝ in actual), exterior, non-loadbearing reinforced concrete masonry wall shown is subjected to 20 lbf/ft² wind pressure. The wind produces a reaction at the roof of 163 lbf/ft and a reaction at the foundation of 117 lbf/ft. The maximum moment produced by the wind is 341 ft-lbf/ft.

compressive strength of masonry	1500 lbf/in²
modular ratio	21.5

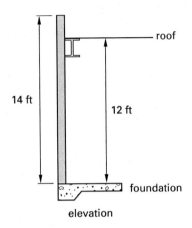

elevation

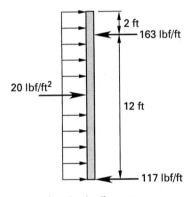

free-body diagram

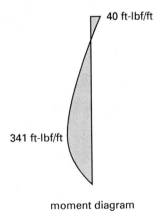

moment diagram

Using a bar spacing of 32 in and grade 60 reinforcing bars, what is most nearly the allowable moment based on the flexural compressive stress in the masonry?

(A) 4100 in-lbf/ft

(B) 4600 in-lbf/ft

(C) 5700 in-lbf/ft

(D) 7600 in-lbf/ft

Hint: Start by determining the required area of reinforcement.

51. A structural masonry wall is designed to span 10 ft vertically from a structure's foundation to the roof. The wall is not reinforced but is grouted solid and carries the load from the roof. The compressive strength of masonry is not specified. Lateral support is provided by intersecting walls spaced 15 ft on center. Using empirical design, what is most nearly the minimum thickness of the wall?

(A) 6 in

(B) 8 in

(C) 9 in

(D) no limit

Hint: A structural masonry wall that supports an axial load (i.e., roof load) is a bearing wall.

52. A wooden pier is supported on kiln-dried round timber piles that extend 20 ft below the water surface into clay soil. A group of three piles carries a 300 kip dead load and a 400 kip live load. The piles are made of red oak and have an area of 230 in². The column stability factor is 0.62. What is the total adjustment factor for the critical compression load case?

(A) 0.551

(B) 0.619

(C) 0.620

(D) 0.688

Hint: Refer to the NDS.

53. A normal-weight (0.150 kip/ft³) reinforced concrete retaining wall is designed to support the loads shown. An 8 ft wide concrete walkway extends along the entire length of the retaining wall. The soil is sandy without silt. The soil behind the retaining wall is soil 1. The soil in front of the retaining wall is soil 2.

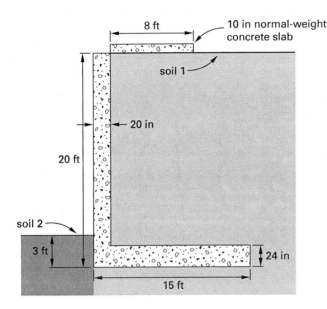

characteristic	soil 1	soil 2
unit weight (kip/ft³)	0.110	0.100
angle of internal friction (degrees)	28	15
cohesion (kip/ft²)	0	0.300
active earth pressure coefficient	0.361	0.589
passive earth pressure coefficient	2.77	1.70

The factor of safety against sliding is

(A) 1.40
(B) 1.44
(C) 1.56
(D) 1.66

Hint: Passive restraint from the soil is only considered if it will always be there. In most cases, it is neglected.

54. A 6.5 ft × 6.5 ft concrete spread footing supports a centrally located 12 in × 12 in concrete column. The dead load on the column, including the column weight, is 50 kips. The live load is 75 kips. The soil report indicates an allowable soil pressure of 4000 lbf/ft². Disregard the weight of the soil. If the concrete compressive strength, f'_c, is 3000 lbf/in² and no. 5 bars are used each way in the footing, the minimum thickness of the footing is most nearly

(A) 6.1 in
(B) 11 in
(C) 12 in
(D) 15 in

Hint: Two-way shear action controls for centrally loaded square footings.

55. A rectangular combined footing is used near a property line to support two 12 in square concrete columns as shown. The allowable soil pressure is 2000 lbf/ft². The compressive strength of concrete is 3000 lbf/in².

column 1
    dead load   30 kips
    live load   60 kips
    moment due to dead load   30 ft-kips
    moment due to live load   40 ft-kips

column 2
    dead load   60 kips
    live load   60 kips
    moment due to dead load   30 ft-kips
    moment due to live load   40 ft-kips

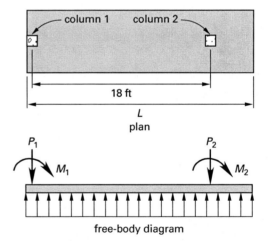

What is most nearly the minimum footing length, L, that will create uniform soil pressure?

(A) 21.6 ft
(B) 21.8 ft
(C) 22.6 ft
(D) 23.0 ft

Hint: For the soil pressure to be uniform, the resultant, R, must be at the centroid of the footing area.

56. Which of the following statements are true?

I. Mat foundations can be used in areas where the basement is below the ground water table (GWT).
II. Mat foundations always require a top layer of reinforcing bars.
III. Conventional spread footings tolerate larger differential settlements than do mat foundations.
IV. Mat foundations are not suitable where settlement may be a problem.

(A) I only
(B) I and II
(C) III and IV
(D) II, III, and IV

Hint: A mat foundation is a large concrete slab in contact with the soil and is commonly used to support several columns or pieces of equipment.

57. An HP12 × 63 steel pile is driven 100 ft into saturated soft clay. The cohesion, c, is 400 lbf/ft^2, and the adhesion, c_A, is 360 lbf/ft^2. What is most nearly the allowable bearing capacity of the pile if the factor of safety is 3?

(A) 1.7 kips
(B) 49 kips
(C) 69 kips
(D) 150 kips

Hint: The allowable bearing capacity is the sum of the point-bearing capacity and the skin-friction capacity divided by the factor of safety.

58. The elevation of an interior girder of a composite two-lane, two-span continuous highway bridge is shown.

steel specification	ASTM A709
minimum web yield stress	50 $kips/in^2$
minimum stiffener yield stress	50 $kips/in^2$
steel modulus of elasticity	29,000 $kips/in^2$

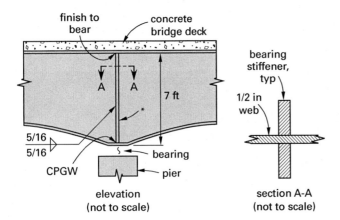

*bearing stiffener 3/4 in × 6 in, typ; each side of web

The factored axial resistance of the effective bearing stiffener column section is most nearly

(A) 570 kips
(B) 590 kips
(C) 650 kips
(D) 730 kips

Hint: Use AASHTO LRFD Bridge Design Specifications (AASHTO) Sec. 6.

Construction Administration

59. A reinforced brick masonry residence located in Los Angeles is designed using strength design provisions. According to *Building Code Requirements for Masonry Structures* (*MSJC*), which of the following requirements must be met during design and construction of this building?

I. Verify placement of reinforcement prior to grouting.
II. Verify placement of grout continuously during construction.
III. Observe preparation of mortar specimens.
IV. Verify the compressive strength of masonry prior to construction and every 5000 ft^2 during construction.

(A) I and II
(B) I and III
(C) I, II, and III
(D) none of the above

Hint: These requirements are part of a quality assurance program.

60. According to the AISC *Steel Construction Manual*, which of the following is NOT an acceptable method of assessing stability requirements of a steel building?

(A) direct analysis method
(B) effective length method
(C) first-order analysis method
(D) plastic drift analysis method

Hint: Refer to Table 2-1 of the AISC *Steel Construction Manual*.

Lateral Breadth Problems

Analysis of Structures

1. A 20 ft diameter, circular water tank 30 ft off the ground provides drinking water to a seaside resort town next to the Pacific Ocean. The basic wind speed for the town is 85 mph. The terrain is flat. The tank is made from concrete, is designed as a rigid structure, and is moderately smooth.

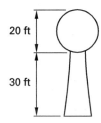

Using the IBC, the design wind force on the tank is most nearly

(A) 25 lbf/ft^2

(B) 3000 lbf

(C) 3500 lbf

(D) 3600 lbf

Hint: The simplified method cannot be used in this case.

2. A retail shopping center located in Utah is built with concrete masonry bearing walls and ordinary reinforced concrete masonry shear walls and is founded on soil. The walls of the shopping center are 15 ft tall. The maximum considered earthquake ground motion for 0.2 sec spectral response acceleration is 30% g. Using the simplified analysis procedure for seismic design found in ASCE7, the seismic base shear is

(A) $0.07W$

(B) $0.14W$

(C) $0.15W$

(D) $14W$

Hint: Refer to ASCE7 Sec. 12.14 for simplified seismic design.

3. Which of the following seismic design statements are true?

I. The seismic base shear of a building increases as the building dead load increases.

II. In areas of high seismic activity, it is best to design buildings with an irregular floor plan to better break up the seismic load.

III. According to ASCE7, flat roof snow loads under 30 lbf/ft^2 need not be included in the effective seismic weight of the structure, W.

IV. Buildings with a soft first story and heavy roofs performed well during the Northridge earthquake in 1994.

(A) I and II

(B) I and III

(C) II and IV

(D) I, II, and III

Hint: Chapter 16 of the IBC provides information on seismic design.

4. A two-story wood-framed apartment building is classified as seismic design category C according to the IBC. Which of the following statements about this building is FALSE?

(A) The total design lateral seismic force increases as the building weight increases.

(B) The short-period response accelerations for this site must be between 0.33 g and 0.50 g.

(C) There is no limit on story drift.

(D) When soil properties are not known in sufficient detail to determine the site class, site class D should be used unless determined otherwise by the building official or unless geotechnical data indicates that site class E or F soil is likely to be present.

Hint: Refer to IBC Sec. 1613.

5. A warehouse facility is earth-bermed to the midpoint of the wall on one side as shown.

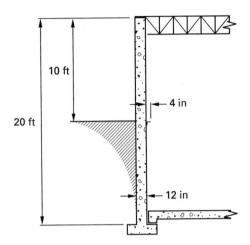

The exterior walls are 12 in reinforced concrete with a compressive strength of 3000 lbf/in². Steel joists framing the roof bear on the full width of the top of the concrete wall. An L6 × 4 × 1/2 (LLV) angle bolted to the concrete wall is used to secure a storage rack and is designed to carry a load of 75 lbf/ft. The first floor is a concrete slab on grade. Assume the wall is simply supported, and disregard the effects of wind. The gravity loads on the wall due to the roof are uniformly distributed and are as follows.

| dead load from roof | 800 lbf/ft |
| live load from roof | 1000 lbf/ft |

If the soil specific weight is 125 lbf/ft³ and K_a is 0.36, the maximum unfactored wall bending moment is most nearly

(A) 4260 ft-lbf/ft

(B) 4350 ft-lbf/ft

(C) 4750 ft-lbf/ft

(D) 4820 ft-lbf/ft

Hint: To determine the maximum moment, consider the eccentricity of the loads.

6. A building is constructed with 9 ft high shear walls as shown.

wall A: 16 in thick, 60 ft long
wall B: 16 in thick, 36 ft long
wall C: 12 in thick, 40 ft long
wall D: 12 in thick, 36 ft long
wall E: 16 in thick, 36 ft long

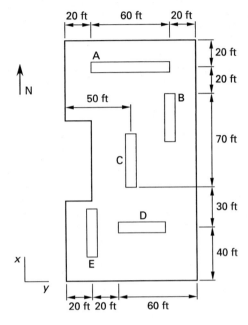

The center of rigidity of the building is located at a point that is most nearly

(A) 115 ft from the south edge of the building and 50.0 ft from the east edge of the building

(B) 115 ft from the south edge of the building and 54.9 ft from the west edge of the building

(C) 123 ft from the north edge of the building and 50.0 ft from the east edge of the building

(D) 123 ft from the south edge of the building and 50.0 ft from the west edge of the building

Hint: The center of rigidity can be based on the relative areas of the walls.

7. A one-story steel-framed building has earth bermed on one side. The total resultant load from the earth pressure is 600 kips and is resisted entirely by the three shear walls as shown. Wall A is 12 in thick. Walls B and C are 14 in thick. The roof is framed with steel joists and corrugated steel deck (no concrete).

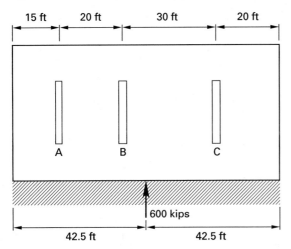

What is most nearly the shear load on wall A?

(A) 124 kips

(B) 176 kips

(C) 180 kips

(D) 247 kips

Hint: A steel joist and corrugated deck roof system is typically considered a flexible diaphragm unless a concrete slab is poured on the roof deck.

8. A building is designed with walls that act in shear only to resist the wind load as shown. The north wind load is 200 lbf/ft. All of the walls are 12 in thick. The center of rigidity (c.r.) of the building is located 98.3 ft from the west facade and 40 ft from the south facade.

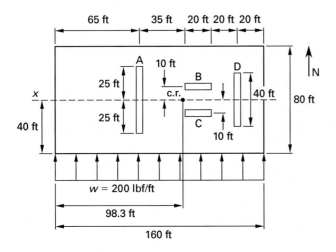

What is most nearly the shear load on wall A?

(A) 6.6 kips

(B) 12 kips

(C) 18 kips

(D) 25 kips

Hint: Wind load is distributed to shear walls according to the relative rigidity of the walls.

9. A loadbearing masonry warehouse is located in central Florida. The building has a flat roof supported by steel joists and metal decking. The roof height is 30 ft plus a 2 ft high parapet. The roof diaphragm is part of the main wind-resisting force system (MWRFS). The building is 40 ft × 90 ft in plan and is classified as enclosed. The structure is designed using allowable stress design. The roof live load is 20 lbf/ft^2, and the roof rain load is 10 lbf/ft^2. The roof dead loads include the following components.

joists	5 lbf/ft^2
roof deck	3 lbf/ft^2
rigid insulation	3 lbf/ft^2
felt and gravel	5 lbf/ft^2

The basic wind speed is 100 mph. Seismic loads do not control. Using the IBC ASD provisions, the critical design load on the roof deck portion of the MWRFS is most nearly

(A) 9.5 lbf/ft^2 (uplift)

(B) 19 lbf/ft^2 (uplift)

(C) 36 lbf/ft^2

(D) 51 lbf/ft^2

Hint: Be sure to consider all load combinations.

10. An enclosed building has a gable roof with a slope of 10°. The mean roof height is 40 ft.

exposure B	
I	1.0
K_{zt}	1.0
basic wind speed	100 mph
effective wind area	20 ft^2

Using the simplified procedure for wind design from ASCE7, the net design wind pressures on the cladding at the corner of the south wall are most nearly

(A) 17 lbf/ft^2, −19 lbf/ft^2

(B) 17 lbf/ft^2, −23 lbf/ft^2

(C) 19 lbf/ft^2, −20 lbf/ft^2

(D) 19 lbf/ft^2, −25 lbf/ft^2

Hint: Use ASCE7 Sec. 6.4.

11. Force F is acting on the frame shown. The framing system is equally spaced between columns. Applying the portal method of frame analysis, what should be the distributed shear be at column II?

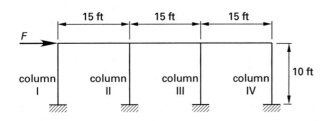

(A) $\frac{1}{8}F$

(B) $\frac{1}{6}F$

(C) $\frac{1}{4}F$

(D) $\frac{1}{3}F$

Hint: Find the value of the shear in each of the columns in terms of the shear in column II.

12. For a flexible roof diaphragm adequately anchored to the shear walls and using a reliability factor of 1.0, which statement identifies the correct chord forces at lines 1 and 2?

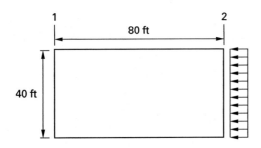

(A) The maximum chord forces at lines 1 and 2 are equal.

(B) The maximum chord force at line 2 is twice the chord force at line 1.

(C) The chord force at line 1 is generally ignored.

(D) The total chord force at lines 1 and 2 is $40w$.

Hint: Determine the relationship between the compression chord force and the tension chord force in a flexible diaphragm.

13. A one-story, wood-frame commercial building has a wood structural panel roof diaphragm, and its south wall has a 40 ft opening. The reliability factor is 1.0. The chord force at the intersection of lines X and 1 is most nearly

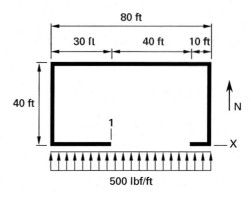

(A) 4700 lbf

(B) 5000 lbf

(C) 9400 lbf

(D) 10,000 lbf

Hint: Use shear and bending moment diagrams across the length of the chord.

Design and Detailing of Structures

14. The reinforced concrete shear wall shown is made from concrete with a compressive strength of 4000 lbf/in^2.

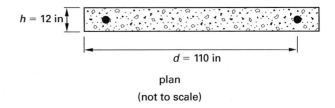

plan
(not to scale)

The nominal shear strength provided by the shear reinforcement is 700 kips. The factored axial gravity load on the wall is 200 lbf/ft. Seismic loads can be ignored. The nominal shear strength of the wall is most nearly

(A) 170 kips

(B) 650 kips

(C) 830 kips

(D) 870 kips

Hint: Refer to ACI 318 Sec. 11.9.

15. A 30 ft high, 10 in thick reinforced concrete shear wall has the following properties.

compressive strength of concrete	6000 lbf/in^2
yield stress of reinforcement	60,000 lbf/in^2
depth to reinforcement	110 in
horizontal length of wall	120 in

The factored in-plane shear force of 110 kips is due to wind loads only. If a single mat of reinforcement is used, the horizontal shear reinforcement required is most nearly

(A) no. 3 at 12 in

(B) no. 3 at 24 in

(C) no. 5 at 12 in

(D) no. 5 at 15 in

Hint: Refer to ACI 318 Sec. 11.9.

16. A 10 ft high, 30 ft long reinforced concrete shear wall has a compressive strength of 4000 lbf/in^2 and a steel reinforcement yield stress of 60,000 lbf/in^2. The ratio of horizontal shear reinforcement area to gross concrete area of vertical section, ρ_t, is 0.0040. The ratio of vertical shear reinforcement area to gross concrete area of horizontal section, ρ_l, should be

(A) 0.0021

(B) 0.0025

(C) 0.0040

(D) 0.0041

Hint: Refer to ACI 318 Sec. 11.9.9.

17. A three-story monolithic reinforced concrete building is supported on columns placed on a grid and spaced 20 ft in each direction. The columns are 18 in × 18 in. Beams with an overall height of 20 in and a web width of 10 in span between the columns in each direction. The slabs are all 6 in thick and are two-way spans. The moment of inertia of the beams for the computation of the relative stiffness parameter, Ψ, at the top of the first-floor corner column is most nearly

(A) 3400 in^4

(B) 3700 in^4

(C) 4100 in^4

(D) 9800 in^4

Hint: The relative stiffness parameter is used in the determination of the effective length factor, k.

18. A 20 in × 20 in brick masonry column with a specified compressive strength of masonry of 3500 lbf/in^2 is reinforced with four no. 6 bars. It carries a compressive load with 5.0 in eccentricity in the x-direction and 6.2 in eccentricity in the y-direction. Compression controls the design.

ratio of effective height to radius of gyration, h/r	72
steel spacing ratio, g	0.4

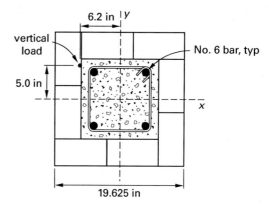

The maximum biaxial load is most nearly

(A) 54,000 lbf

(B) 96,000 lbf

(C) 100,000 lbf

(D) 250,000 lbf

Hint: The maximum biaxial load is not a linear sum of the individual capacities.

19. A continuous L4 × 6 × $^3/_8$ angle anchored to a 12 in concrete masonry wall is used as a ledger for floor joists as shown. A307 headed anchor bolts are placed 16 in on center in grouted cells. The joists are spaced 6 ft on center. The specified compressive strength of masonry is 1500 lbf/in^2.

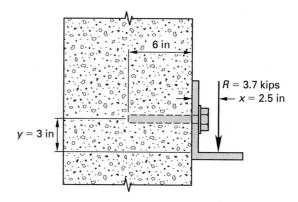

If the load from the joists is 3.7 kips, what is the smallest diameter bolt that can be used?

(A) $^1/_4$ in

(B) $^3/_8$ in

(C) $^5/_8$ in

(D) $^3/_4$ in

Hint: The bolt is subject to both shear and tension.

20. The cantilevered sheet piling shown has a concentrated lateral load, H, of 10 kips spaced 2 ft on center, horizontally along the top of the piling. The soil specific weight is 110 lbf/ft^3, and the angle of internal friction of the soil is 30°.

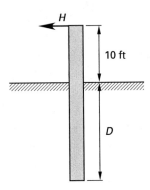

If the factor of safety for the minimum depth is 1.3, the total length of the sheet pile is most nearly

(A) 21 ft

(B) 26 ft

(C) 31 ft

(D) 38 ft

Hint: The solution for this problem is an iterative one.

21. A three-story structural steel-framed office building is classified as a seismic design category B structure. The floor-to-floor height is 12 ft, and the structure is rectangular in plan. The interior partitions, ceilings, and walls have not been designed to accommodate story drifts. According to ASCE7, the maximum design story drift at the second story is

(A) 0.24 in

(B) 2.2 in

(C) 2.9 in

(D) 3.6 in

Hint: Refer to ASCE7 Sec. 12.12.

22. A 10 ft high, 20 ft long special reinforced masonry shear wall is subjected to in-plane seismic loading. The wall is constructed from 12 in concrete masonry units laid in running bond and type S mortar. All reinforcement is adequate to resist the applied shear loads. Which shear reinforcement most nearly complies with the minimum reinforcement requirements?

(A) one no. 5 bar in horizontal bond beams, spaced 32 in on center

(B) one no. 5 bar in horizontal bond beams, spaced 32 in on center, and no. 5 vertical bars, spaced at 32 in on center

(C) two no. 4 bars in horizontal bond beams, spaced 48 in on center, and no. 5 vertical bars, spaced at 32 in on center

(D) two no. 5 bars in horizontal bond beams, spaced 24 in on center, and no. 6 vertical bars, spaced 16 in on center

Hint: Use *Building Code Requirements for Masonry Structures* (MSJC) Sec. 1.17.3.2.

23. Which of the following is NOT a requirement of a structural steel special moment-resisting frame (SMF)?

(A) Beam-to-column connections used in a seismic load-resisting system shall be capable of sustaining an interstory drift angle of at least 0.02 rad.

(B) Beam-to-column connections must be prequalified for SMF in accordance with AISC 341 App. P.

(C) Both flanges of beams shall be laterally braced.

(D) The maximum beam flange bracing spacing is $0.086 r_y E/F_y$.

Hint: Refer to ASCE7 Chap. 14.

24. A wood-frame shear wall is framed with 2 × 4 studs at 16 in on center. The exterior of the wall has $3/8$ in OSB sheathing attached with 8d nails, and the panel edge fastener spacing is 6 in. The interior is sheathed with $1/2$ in gypsum wallboard attached with no. 6 type S drywall screws ($1^{1}/_{4}$ in long) spaced at 8 in around the edges. The nominal unit seismic shear capacity of the wall is most nearly

(A) 240 lbf/linear ft

(B) 440 lbf/linear ft

(C) 580 lbf/linear ft

(D) 600 lbf/linear ft

Hint: Refer to *Special Design Provisions for Wind and Seismic* (SDPWS).

25. Which statement(s) regarding *Special Design Provisions for Wind and Seismic* (SDPWS) is/are true?

I. Wood-framed shear walls sheathed with gypsum wallboard are permitted to resist seismic forces in seismic design category C.

II. Wood-framed shear walls sheathed with gypsum wallboard can be constructed as blocked or unblocked.

III. Wood-frame shear walls with particleboard sheathing are permitted to resist seimic forces in seismic design category D.

IV. Single-layer lumber used to diagonally sheathe wood-frame shear walls must have a nominal thickness of at least 1 in.

(A) I only

(B) II and III

(C) I, II, and IV

(D) II, III, and IV

Hint: Refer to SDPWS Sec. 4.3.

26. The plan and elevation views of a reinforced concrete bridge pier is shown.

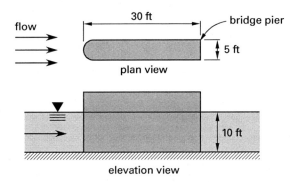

If the water velocity of the creek's design flood is 10 ft/sec, the unfactored longitudinal drag force acting on the upstream edge of the bridge pier is most nearly

(A) 0.35 kips

(B) 3.5 kips

(C) 7.0 kips

(D) 21 kips

Hint: Use *AASHTO LRFD Bridge Design Specifications* (AASHTO) Sec. 3.7.

27. A loadbearing concrete masonry wall in a large elementary school is 12 ft high × 30 ft long. The wall weighs 1350 lbf/linear ft and supports a precast concrete plank floor weighing 40 lbf/ft². The design earthquake spectral response acceleration parameter at short periods, S_{DS}, is 0.46.

The anchorage of the wall to the precast concrete plank flooring must be capable of resisting a loading that is most nearly

(A) 248 lbf/linear ft

(B) 280 lbf/linear ft

(C) 311 lbf/linear ft

(D) 320 lbf/linear ft

Hint: Use ASCE7 Sec. 12.11.

Construction Administration

28. Structural observation must be provided for a seismic design category D structure that is

(A) a two-story office building

(B) a four-story office building

(C) a public high school

(D) an agricultural arena with a height of 60 ft

Hint: Refer to IBC Sec. 1710.2.

29. Which statement is INCORRECT according to OSHA safety management regulations?

(A) An employer must use OSHA forms 300, 300A, and 301 to record workplace injuries and illnesses.

(B) An employer does not have to keep safety records if it has fewer than 10 employees at all times during a calendar year.

(C) An employer must record motor vehicle accidents of employees commuting to or from work.

(D) An employer must orally report the death of an employee from a work-related accident to an OSHA office within eight hours of the accident.

Hint: Consult OSHA Std. 1904.

30. Half of a shipment of steel reinforcements arrives at a project site lightly coated with steel mill scale and mud. The construction manager should

(A) reject the entire shipment from the reinforcement supplier

(B) reject only the reinforcements coated with mill scale and mud

(C) sand blast the surfaces of the coated reinforcements before placing them in formwork

(D) do nothing because the reinforcements are fine as they are

Hint: Consult ACI 318 Chap. 7.

Solutions
Vertical Breadth Problems

1. To determine what loads need to be considered on a lintel, first determine if arching action will occur. A masonry wall will exhibit arching action over an opening if sufficient masonry extends on both sides of the opening to resist the thrusting action of the arch. An extent of masonry on each side greater than the distance of the opening itself is usually adequate to resist the thrust. When arching action occurs, only the loads within a 45° triangle over the opening will be carried by the lintel.

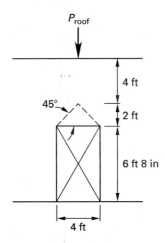

The concentrated load is distributed downward at an angle of 45° over an effective length equal to four times the wall thickness plus the width of the bearing [*Building Code Requirements for Masonry Structures (MSJC)* Sec. 2.1.9]. In this case, the distributed concentrated load from the roof truss occurs beyond the apex of the 45° triangle and therefore is not included in the loads on the lintel.

Calculate the loads on the lintel.

The span length of the lintel is the center-of-bearing to center-of-bearing distance—the clear span plus one-half the bearing length on each side of the opening. *MSJC* Sec. 2.3.3.4 specifies a minimum bearing length of 4 in for beams. The span length is

$$L = \text{clear span} + \tfrac{1}{2}\sum \text{bearing length at each end}$$
$$= 4 \text{ ft} + \left(\tfrac{1}{2}\right)(4 \text{ in} + 4 \text{ in})\left(\frac{1 \text{ ft}}{12 \text{ in}}\right)$$
$$= 4.33 \text{ ft}$$

Determine the lintel self-weight. From the AISC *Steel Construction Manual*'s Table of Dimensions and Properties, the weight of the two $5 \times 3^{1}/_{2} \times {}^{5}/_{16}$ angles is

$$w_{\text{lintel}} = (2)\left(8.7 \, \frac{\text{lbf}}{\text{ft}}\right) = 17.4 \text{ lbf/ft}$$

The weight of the masonry above the lintel is a triangular load. At its maximum, this load is

$$w_{\text{peak}} = w_{\text{uniform}}\left(\frac{L}{2}\right)$$
$$w_{\text{uniform}} = t\gamma = (7.625 \text{ in})\left(\frac{1 \text{ ft}}{12 \text{ in}}\right)\left(150 \, \frac{\text{lbf}}{\text{ft}^3}\right)$$
$$= 95 \text{ lbf/ft}^2$$
$$w_{\text{peak}} = \left(95 \, \frac{\text{lbf}}{\text{ft}^2}\right)\left(\frac{4.33 \text{ ft}}{2}\right) = 205.7 \text{ lbf/ft}$$

The total triangular load is

$$W = \tfrac{1}{2}Lw_{\text{peak}} = \left(\tfrac{1}{2}\right)(4.33 \text{ ft})\left(205.7 \, \frac{\text{lbf}}{\text{ft}}\right)$$
$$= 445 \text{ lbf}$$

The maximum moment on the lintel is

$$M = \frac{w_{\text{lintel}}L^2}{8} + \frac{WL}{6}$$
$$= \frac{\left(17.4 \, \frac{\text{lbf}}{\text{ft}}\right)(4.33 \text{ ft})^2}{8} + \frac{(455 \text{ lbf})(4.33 \text{ ft})}{6}$$
$$= 369 \text{ ft-lbf} \quad (400 \text{ ft-lbf})$$

The answer is (B).

Why Other Options Are Wrong

(A) This incorrect solution uses the clear span for the span length of the lintel.

(C) This incorrect solution uses the clear span for the span length of the lintel and, although it correctly calculates the triangular load, it also mistakenly includes the load from the roof truss in the load on the lintel.

(D) This incorrect solution correctly calculates the triangular load on the lintel but mistakenly includes the load from the roof truss.

2. The load on the beam, w, is the weight of the wall per foot.

$$w = \gamma h t = \left(85 \; \frac{\text{lbf}}{\text{ft}^3}\right)(8 \text{ ft})(6 \text{ in})\left(\frac{1 \text{ ft}}{12 \text{ in}}\right)$$
$$= 340 \text{ lbf/ft}$$

The answer is (A).

Why Other Options Are Wrong

(B) This incorrect solution results from failing to multiply by the wall width.

(C) This incorrect solution is a result of calculating the total load rather than the load per foot.

(D) This incorrect solution fails to perform the conversion from inches to feet.

3. Section 1608 of the IBC covers snow loads and states that design snow loads shall be determined in accordance with ASCE7 Sec. 7. Begin by determining the flat roof snow load and adjusting it for the roof slope.

The flat roof snow load is

$$p_f = 0.7 C_e C_t I p_g \quad \text{[ASCE7 Eq. 7-1]}$$

The snow exposure factor, C_e, is given in ASCE7 Table 7-2 based on the terrain exposure category and roof exposure category. The terrain exposure category given in ASCE7 Sec. 6.5.6 for an office park is exposure B. The roof exposure category for a suburban location can be assumed to be partially exposed. Based on these factors, C_e is 1.0.

The thermal factor, C_t, is given in ASCE7 Table 7-3 as 1.1 for ventilated attics with insulation greater than R-25.

The importance factor, I, is given in ASCE7 Table 7-4. An office building falls into category II. I is 1.0.

$$p_f = 0.7 C_e C_t I p_g$$
$$= (0.7)(1.0)(1.1)(1.0)\left(40 \; \frac{\text{lbf}}{\text{ft}^2}\right)$$
$$= 30.8 \text{ lbf/ft}^2$$

Check the minimum low slope roof snow load as specified in ASCE7 Sec. 7.3. For ground snow loads over 20 lbf/ft²,

$$p_{\min} = I\left(20 \; \frac{\text{lbf}}{\text{ft}^2}\right) = (1.0)\left(20 \; \frac{\text{lbf}}{\text{ft}^2}\right)$$
$$= 20 \text{ lbf/ft}^2 \quad [< 30.8 \text{ lbf/ft}^2, \text{OK}]$$

Use ASCE7 Sec. 7.4 to determine the sloped roof snow load, p_s.

$$p_s = C_s p_f \quad \text{[ASCE7 Eq. 7-2]}$$

The roof slope factor, C_s, for cold roofs is found in ASCE7 Sec. 7.4.2. For a C_t of 1.1, C_s is determined from ASCE7 Fig. 7-2b using the solid line for an asphalt shingle roof [ASCE7 Sec. 7.4]. For a roof pitch of 6:12, determine that

$$C_s = 1.0 \quad \text{[ASCE7 Eq. 7-2b]}$$

Therefore, the sloped roof snow load is

$$p_s = C_s p_f = (1.0)\left(30.8 \; \frac{\text{lbf}}{\text{ft}^2}\right) = 30.8 \text{ lbf/ft}^2$$

To determine the maximum snow load, the unbalanced condition must also be considered as found in ASCE7 Sec. 7.6. For hip and gable roofs, W is the horizontal eave-to-ridge distance, or 20 ft. According to ASCE7 Sec. 7.6.1, if W is 20 ft or less, the unbalanced snow load on the leeward side is

$$I p_g = (1.0)\left(40 \; \frac{\text{lbf}}{\text{ft}^2}\right) = 40 \text{ lbf/ft}^2$$

The unbalanced snow load on the windward side is zero.

The unbalanced load condition is greater than the balanced load condition. The maximum snow load on the leeward roof is 40 lbf/ft².

The answer is (C).

Why Other Options Are Wrong

(A) This incorrect solution finds the sloped roof snow load and ignores the unbalanced condition.

(B) This incorrect solution determines the roof slope factor for a slippery surface.

(D) This incorrect solution uses the 2002 ASCE7 to determine the unbalanced snow load.

4. The induced axial load in a constrained member due to a temperature change is

$$P_{\text{th}} = \alpha(T_2 - T_1)AE$$

α is the coefficient of thermal expansion of the member. According to Chap. 2 of the AISC *Steel Construction Manual*, the coefficient of thermal expansion for mild steel at temperatures up to 100°F is

$$\alpha = \frac{0.00065}{100°\text{F}}$$

The area of the steel bar is

$$A = \pi\left(\frac{d^2}{4}\right) = \pi\left(\frac{(2.5 \text{ in})^2}{4}\right)$$
$$= 4.91 \text{ in}^2$$
$$P_{\text{th}} = \alpha(T_2 - T_1)AE$$
$$= \left(\frac{0.00065}{100°\text{F}}\right)(100°\text{F} - 35°\text{F})(4.91 \text{ in}^2)$$
$$\quad \times \left(29 \times 10^6 \; \frac{\text{lbf}}{\text{in}^2}\right)\left(\frac{1 \text{ kip}}{1000 \text{ lbf}}\right)$$
$$= 60.2 \text{ kips} \quad (60 \text{ kips})$$

The answer is (A).

Why Other Options Are Wrong

(B) This incorrect solution uses the maximum temperature instead of the temperature change to calculate the induced axial load.

(C) This solution incorrectly calculates the area of the bar as πd^2 instead of $\pi d^2/4$.

(D) This incorrect solution reads the coefficient of thermal expansion as having no units instead of as 0.00065 per 100°F. The units do not work out in the axial load equation.

5. Section 1807.1.6 of the IBC includes prescriptive requirements for concrete and masonry foundation walls.

Section 1807.1.6.1 of the IBC specifies that the minimum thickness of foundation walls must not be less than the thickness of the wall supported. Therefore, the minimum thickness of the foundation wall must be equal to the concrete masonry wall thickness, or 11.63 in.

Verify that 11.63 in is adequate based on the soil lateral load. IBC Sec. 1610 requires that foundation walls that are laterally supported at the top be designed for at-rest pressure. Using IBC Table 1610.1, poorly graded clean sands and sand-gravel mixes have a unified soil classification of SP and an at-rest design lateral soil load of 60 lbf/ft²-ft.

IBC Table 1807.1.6.2 specifies the minimum thickness for concrete foundation walls. The minimum thickness depends upon the wall height and the height of the unbalanced backfill. The wall height is 8 ft. The height of unbalanced backfill is 8 ft minus 4 ft, which equals 4 ft. Knowing these values and the design lateral soil load of 60 lbf/ft²-ft, IBC Table 1807.1.6.2 requires a minimum wall thickness of 7.5 in.

The answer is (D).

Why Other Options Are Wrong

(A) This incorrect solution overlooks the requirements of IBC Sec. 1807.1.6.1 for minimum thickness equal to the supported wall thickness.

(B) This incorrect solution overlooks the requirements of IBC Sec. 1807.1.6.1 for minimum thickness equal to the supported wall thickness and uses IBC Table 1807.1.6.3(1) for plain masonry walls instead of plain concrete walls.

(C) This incorrect solution miscalculates the height of unbalanced backfill in IBC Table 1807.1.6.2 as 8 ft instead of 4 ft and overlooks the requirements of IBC Sec. 1807.1.6.1 for minimum thickness equal to the supported wall thickness.

6. Section 13.2 of ACI 318 defines a column strip as a width on each side of a column centerline equal to $0.25l_2$ or $0.25l_1$, whichever is less. A middle strip is the strip bound by two column strips.

l_1 is defined as the length of span in the direction in which moments are being determined, measured center-to-center of the supports.

l_2 is the length of span transverse to l_1, measured center-to-center of the supports.

l_n is the length of clear span in the direction that moments are being determined, measured face-to-face of the supports.

The midspan moment of the middle strip will be a positive moment. To find the moment in the middle strip, first determine the column strip moment.

Determine the factored load on the slab [ACI 318 Sec. 9.2].

$$w_u = 1.2D + 1.6L = (1.2)\left(100 \; \frac{\text{lbf}}{\text{ft}^2}\right) + (1.6)\left(30 \; \frac{\text{lbf}}{\text{ft}^2}\right)$$
$$= 168 \text{ lbf/ft}^2$$

The total factored static moment on the slab given in ACI 318 Sec. 13.6.2.2 is

$$l_1 = 40 \text{ ft}$$
$$l_2 = 30 \text{ ft}$$
$$l_n = 40 \text{ ft} - (2)\left(\frac{1 \text{ ft}}{2}\right) = 39 \text{ ft}$$
$$M_o = \frac{w_u l_2 l_n^2}{8} = \frac{\left(168 \; \frac{\text{lbf}}{\text{ft}^2}\right)\left(\frac{1 \text{ kip}}{1000 \text{ lbf}}\right)(30 \text{ ft})(39 \text{ ft})^2}{8}$$
$$= 958 \text{ ft-kips}$$

The total static moment, M_o, is distributed as

$$-M_u = -0.65 M_o$$
$$M_u = 0.35 M_o$$

Since the midspan moment is positive, use the factored positive moment to find the positive moment in the column and middle strips.

$$M_u = 0.35 M_o = (0.35)(958 \text{ ft-kips}) = 335 \text{ ft-kips}$$

ACI 318 Sec. 13.6.4.4 gives the percentage of positive factored moment distributed to the column strips based on the ratio of the stiffness of the beam to the stiffness of the slab, $\alpha_{f1} l_2 / l_1$. Since this is a flat plate, there are no beams and α_{f1} is zero.

$$\frac{\alpha_{f1} l_2}{l_1} = \frac{(0)(30 \text{ ft})}{40 \text{ ft}} = 0$$

Use the table in ACI 318 Sec. 13.6.4.4 to find the percentage of positive factored moment distributed to the column strips to be 60%.

Since the slab is symmetric, the percentage of positive moment in the middle strip is

$$M_{\text{middle},\%} = 100\% - 60\% = 40\%$$

The positive moment in the middle strip is

$$M_{\text{middle}} = (40\%) M_u = (0.40)(335 \text{ ft-kips})$$
$$= 134 \text{ ft-kips} \quad (130 \text{ ft-kips})$$

The answer is (C).

Why Other Options Are Wrong

(A) This incorrect solution uses α_{f1} equal to 1.0 instead of 0 when calculating the percentage of positive moment distributed to the column strips.

(B) This incorrect solution reverses the values of l_1 and l_2 in calculating the total factored static moment on the slab, M_o. l_1 is the direction in which the moments are being determined.

(D) This incorrect solution uses the center-to-center span length instead of the clear span, l_n, in determining the total factored static moment.

7. From the sum of the moments about E, the vertical reaction at support A is

$$R_{A,v} = \frac{(300 \text{ lbf})(2h) + (200 \text{ lbf})(3h)}{4h} = 300 \text{ lbf}$$

Draw the free-body diagram of joint A.

By summing the forces in the horizontal direction, determine that the horizontal reaction at A is zero.

The sum of the forces in the vertical direction is

$$AH_v - R_{A,v} = 0 \text{ lbf}$$
$$AH_v = 300 \text{ lbf} \text{ (compression)}$$

From the geometry of the truss, the horizontal component of AH must be twice the vertical component.

$$AH_h = (2)(300 \text{ lbf}) = 600 \text{ lbf} \text{ (compression)}$$

The resultant force in member AH is

$$AH = \sqrt{AH_v^2 + AH_h^2} = \sqrt{(300 \text{ lbf})^2 + (600 \text{ lbf})^2}$$
$$= 670.8 \text{ lbf} \quad (670 \text{ lbf} \text{ (compression)})$$

The answer is (C).

Why Other Options Are Wrong

(A) This incorrect solution finds only the vertical component of the force in AH.

(B) This incorrect solution finds the horizontal component of the force in AH.

(D) This incorrect solution identifies the resultant force in AH as a tensile force.

8. The degree of indeterminacy of a pin-connected truss is given by the equation

$$\text{degree of indeterminacy} = 3 + \text{number of members}$$
$$- 2(\text{number of joints})$$

In this case, there is one degree of indeterminacy or redundancy.

$$\text{degree of indeterminacy} = 3 + 6 - (2)(4) = 1$$

The forces in an indeterminate truss cannot be solved directly. Since there is only one redundant member, use the dummy unit-load method to determine the force in member BC.

step 1: Draw the truss twice. Omit the redundant member on both trusses.

step 2: Load the first truss (which is now determinate) with the actual loads.

step 3: Calculate the forces, S, in all of the members. Use a positive sign for tensile forces.

step 4: Load the second truss with two unit forces acting collinearly toward each other along the line of the redundant member.

step 5: Calculate the force, u, in each of the members.

step 6: Calculate the force in the redundant member using the equation

$$S_{\text{redundant}} = \frac{-\sum \dfrac{SuL}{AE}}{\sum \dfrac{u^2 L}{AE}}$$

step 7: The true force in member j of the truss is

$$F_{j,\text{true}} = S_j + S_{\text{redundant}} u_j$$

Removing the redundant member, draw the free-body diagram of the loaded truss. The reaction loads are calculated by static analysis of the truss.

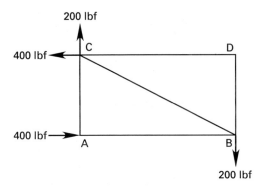

Draw the free-body diagram for each joint to determine the force in the members.

Joint A:

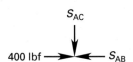

Summing the forces,

$$S_{AC} = 0$$
$$S_{AB} = -400 \text{ lbf}$$

Joint B:

Summing the forces, the horizontal component of the force in member BC must equal 400 lbf (tension). By geometry of the figure, determine that the vertical component of member BC equals 200 lbf. Therefore, the force in member BD is zero. The resultant force in member BC is

$$S_{BC} = \sqrt{(400 \text{ lbf})^2 + (200 \text{ lbf})^2}$$
$$= 447 \text{ lbf}$$

Continue in the same fashion, solving for the forces in the truss.

member	L (ft)	AE (kips)	S (lbf)	u	$\dfrac{SuL}{AE}$ (lbf-ft/kip)	$\dfrac{u^2 L}{AE}$ (ft/kip)
AB	10	3	-400	-0.894	1192	2.66
AC	5	3	0	-0.447	0	0.33
CD	10	3	0	-0.894	0	2.66
BC	11.18	5	447	1.0	999	2.24
BD	5	3	0	-0.447	0	0.33
AD	11.18	5	0	1.0	0	2.24
					2191	10.46

Draw the free-body diagram of the unit-load truss, and solve for the member forces.

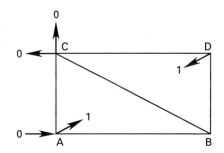

The force in redundant member AD is

$$S_{AD} = \frac{-\sum \dfrac{SuL}{AE}}{\sum \dfrac{u^2 L}{AE}} = \frac{-2191 \dfrac{\text{lbf-ft}}{\text{kip}}}{10.46 \dfrac{\text{ft}}{\text{kip}}}$$
$$= -209 \text{ lbf}$$

The true force in member BC of the truss is

$$F_{BC,\text{true}} = S_{BC} + S_{\text{redundant}} u_{BC}$$
$$= 447 \text{ lbf} + (-209 \text{ lbf})(1.0)$$
$$= 238 \text{ lbf (tension)} \quad (240 \text{ lbf (tension)})$$

The answer is (C).

Why Other Options Are Wrong

(A) This incorrect solution calculates the force in redundant member AD instead of the force in member BC.

(B) This incorrect solution uses the same product of area and modulus of elasticity, AE, for all members of the truss.

(D) This incorrect solution calculates the force in member BC in the determinate truss due to the applied load, S_{BC}, instead of the true force in member BC, $F_{BC,\text{true}}$.

9. To find the centroid of the cross section, divide the section into two rectangles.

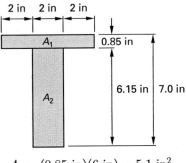

$$A_1 = (0.85 \text{ in})(6 \text{ in}) = 5.1 \text{ in}^2$$
$$A_2 = (6.15 \text{ in})(2 \text{ in}) = 12.3 \text{ in}^2$$

The distance from the top of the section to the centroid is

$$y_c = \frac{\sum A_i y_{ci}}{\sum A_i}$$
$$= \frac{(5.1 \text{ in}^2)\left(\frac{0.85 \text{ in}}{2}\right) + (12.3 \text{ in}^2)\left(\frac{6.15 \text{ in}}{2} + 0.85 \text{ in}\right)}{5.1 \text{ in}^2 + 12.3 \text{ in}^2}$$
$$= 2.9 \text{ in} \quad \text{[from top of section]}$$

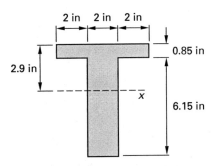

The shear stress is

$$\tau = \frac{VQ}{Ib}$$

Calculate I and Q. For a rectangular section,

$$I_1 = \frac{bh^3}{12} = \frac{(6 \text{ in})(0.85 \text{ in})^3}{12} = 0.31 \text{ in}^4$$
$$I_2 = \frac{bh^3}{12} = \frac{(2 \text{ in})(6.15 \text{ in})^3}{12} = 38.8 \text{ in}^4$$

The moment of inertia about the centroid is

$$I_x = \sum(I_i + A_i d_i^2)$$
$$= 0.31 \text{ in}^4 + (5.1 \text{ in}^2)\left(2.9 \text{ in} - \frac{0.85 \text{ in}}{2}\right)^2 + 38.8 \text{ in}^4$$
$$+ (12.3 \text{ in}^2)\left(7.0 \text{ in} - \frac{6.15 \text{ in}}{2} - 2.9 \text{ in}\right)^2$$
$$= 83.3 \text{ in}^4$$

The statical moment of the area, Q, is the product of the area above or below the point in question and the distance from the centroidal axis to the centroid of the area. Looking at the area below the centroid,

$$Q = A\bar{y} = \tfrac{1}{2}b(h - y_c)^2$$
$$= \left(\tfrac{1}{2}\right)(2 \text{ in})(7.0 \text{ in} - 2.9 \text{ in})^2$$
$$= 16.8 \text{ in}^3$$

The shear stress at the centroid is

$$\tau = \frac{VQ}{Ib} = \frac{(100 \text{ lbf})(16.8 \text{ in}^3)}{(83.3 \text{ in}^4)(2 \text{ in})}$$
$$= 10.1 \text{ lbf/in}^2 \quad (10 \text{ lbf/in}^2)$$

The answer is (C).

Why Other Options Are Wrong

(A) This incorrect solution miscalculates the location of the centroid of the section. The distance from the centroid of area A_2 to the top of the section fails to include the 0.85 in thickness of area A_1.

(B) This incorrect solution miscalculates A_2 and carries the mistake throughout the subsequent calculations. The length of A_2 is taken as the overall length of 7.0 in instead of the actual length of 6.15 in.

(D) This incorrect solution does not properly calculate the transformed moment of inertia. It directly adds the moments of inertia for each area instead of calculating the transformed moment of inertia.

10. Moment distribution is based on the relative stiffness of the members.

The relative stiffness of each section is given as

$$K_{AB} = 0.286/\text{ft}$$
$$K_{BC} = 0.190/\text{ft}$$

The distribution factor for a joint is

$$DF = \frac{K}{\sum K}$$

$$DF_{AB} = \frac{K_{AB}}{K_{AB}} = \frac{\frac{0.286}{ft}}{\frac{0.286}{ft}} = 1.0$$

$$DF_{BA} = \frac{K_{AB}}{K_{AB} + K_{BC}} = \frac{\frac{0.286}{ft}}{\frac{0.286}{ft} + \frac{0.190}{ft}} = 0.6$$

$$DF_{BC} = \frac{K_{BC}}{K_{AB} + K_{BC}} = \frac{\frac{0.190}{ft}}{\frac{0.286}{ft} + \frac{0.190}{ft}} = 0.4$$

$$DF_{CB} = 0 \quad \text{[fixed end]}$$

The fixed-end moments (FEM), as taken from a reference text, are

$$FEM_{AB} = -\frac{Pb^2 a}{L^2} = -\frac{(100 \text{ kips})(8 \text{ ft})^2 (6 \text{ ft})}{(14 \text{ ft})^2}$$
$$= -195.9 \text{ ft-kips}$$

$$FEM_{BA} = \frac{Pa^2 b}{L^2} = \frac{(100 \text{ kips})(6 \text{ ft})^2 (8 \text{ ft})}{(14 \text{ ft})^2}$$
$$= 146.9 \text{ ft-kips}$$

$$FEM_{BC} = -\frac{wL^2}{30} = -\frac{\left(50 \frac{\text{kips}}{\text{ft}}\right)(21 \text{ ft})^2}{30}$$
$$= -735.0 \text{ ft-kips}$$

$$FEM_{CB} = \frac{wL^2}{20} = \frac{\left(50 \frac{\text{kips}}{\text{ft}}\right)(21 \text{ ft})^2}{20}$$
$$= 1102.5 \text{ ft-kips}$$

Use moment distribution to find the moments at the supports.

Start the moment distribution at joint B. The unbalanced moment of -588.1 ft-kips is reversed in sign and multiplied by the distribution factors, 0.6 and 0.4. This puts a balancing correction of 352.9 ft-kips on the left side of joint B and 235.2 ft-kips on the right side of joint B, for a total balancing correction of 588.1 ft-kips. (The horizontal lines under the moments of 352.09 ft-kips and 235.2 ft-kips indicated that joint B has now been balanced.) Multiply the balancing moments by the carryover factors of 0.5, and carry the products over to the far ends of AB and BC.

Continue by moving to joint A and balancing it in a similar manner. Continue until the carryover values are small in magnitude. End when joint B is balanced.

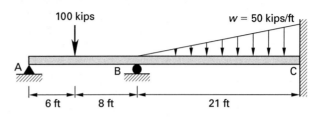

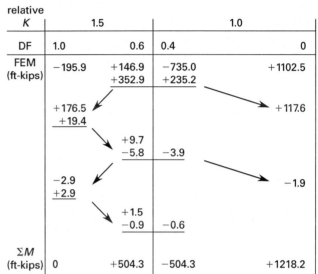

The end moments are obtained by summing all the entries at each location.

The moment at the fixed end is

$$M_C = 1102.5 \text{ ft-kips} + 117.6 \text{ ft-kips} - 1.9 \text{ ft-kips}$$
$$= 1218.2 \text{ ft-kips} \quad (1200 \text{ ft-kips})$$

As a check, the moment on each side of joint B should sum to zero, and the moment at the pinned end should sum to zero.

A positive moment indicates clockwise rotation at the joint, according to the sign convention used.

The answer is (B).

Why Other Options Are Wrong

(A) This incorrect solution reverses the FEMs for the triangular load. The moments on each side of joint B should be equal. In this case, they do not work out to be equal.

(C) This incorrect solution solves the FEM distribution correctly as in the solution. However, the sign convention is not applied correctly, and a counterclockwise rotation is assumed.

(D) This incorrect solution reverses the distribution factors at joint B, putting the distribution factor for BA at BC and vice versa.

11. Find the reactions.

Member AB:

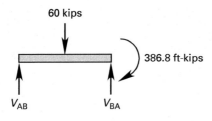

Taking clockwise moments and upward forces as positive,

$$\sum M_A = 0 + (60 \text{ kips})(10 \text{ ft})$$
$$+ 386.8 \text{ ft-kips} - V_{BA}(20 \text{ ft}) = 0$$
$$V_{BA} = 49.3 \text{ kips}$$

Member BC:

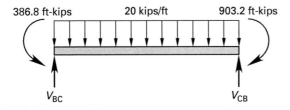

Taking clockwise moments and upward forces as positive,

$$\sum M_C = 903.2 \text{ ft-kips} - \left(20 \ \frac{\text{kips}}{\text{ft}}\right)(20 \text{ ft})(10 \text{ ft})$$
$$- 386.8 \text{ ft-kips} + V_{BC}(20 \text{ ft})$$
$$= 0$$
$$V_{BC} = 174.2 \text{ kips}$$

The reaction at B is the sum of the shears at B.

$$R_B = V_{BA} + V_{BC} = 49.3 \text{ kips} + 174.2 \text{ kips}$$
$$= 223.5 \text{ kips} \quad (220 \text{ kips})$$

The answer is (D).

Why Other Options Are Wrong

(A) This incorrect solution only considers the shear to the left of B in determining the reaction at B instead of including the shear from BC.

(B) This incorrect solution subtracts the shear forces on each side of B rather than adding them to determine the reaction at the support.

(C) This incorrect solution only considers the shear to the right of B in determining the reaction at B instead of including the shear from BA as well.

12. To determine deflection by the conjugate beam method, draw the M/EI diagram for the actual load, then load the conjugate beam with the M/EI diagram. The deflection of the beam at a point is numerically equal to the moment of the conjugate beam at that point.

Draw the free-body diagram for the beam.

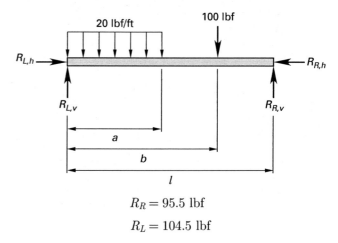

$$R_R = 95.5 \text{ lbf}$$
$$R_L = 104.5 \text{ lbf}$$

Draw the moment diagram for the beam with the actual loads.

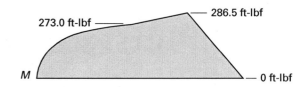

The moment at 5 ft from the left support is

$$M_{5 \text{ ft}} = R_R(6 \text{ ft}) - (100 \text{ lbf})(3 \text{ ft})$$
$$= (95.5 \text{ lbf})(6 \text{ ft}) - (100 \text{ lbf})(3 \text{ ft})$$
$$= 273.0 \text{ ft-lbf}$$

The moment at 8 ft from the left support is

$$M_{8 \text{ ft}} = R_R(3 \text{ ft}) = (95.5 \text{ lbf})(3 \text{ ft})$$
$$= 286.5 \text{ ft-lbf}$$

To determine the deflection, draw the conjugate beam loaded with the M/EI diagram.

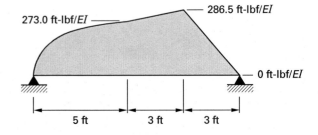

Find the equivalent loads by dividing the moment diagram into easy-to-calculate areas. The area under the curve can be approximated as the area of a parabola.

$$P_{\text{parabola}} = \tfrac{2}{3}hb = \left(\tfrac{2}{3}\right)\left(\frac{273.0 \text{ ft-lbf}}{EI}\right)(5 \text{ ft})$$
$$= 910 \text{ lbf-ft}^2/EI$$

$$P_{\text{rectangle}} = bh = (3 \text{ ft})\left(\frac{273.0 \text{ ft-lbf}}{EI}\right)$$
$$= 819 \text{ lbf-ft}^2/EI$$

$$P_{\text{triangle}} = \tfrac{1}{2}hb = \left(\tfrac{1}{2}\right)\left(\frac{286.5 \text{ ft-lbf} - 273.0 \text{ ft-lbf}}{EI}\right)(3 \text{ ft})$$
$$= 20.25 \text{ lbf-ft}^2/EI$$

$$P_{\text{triangle}} = \tfrac{1}{2}hb = \left(\tfrac{1}{2}\right)\left(\frac{286.5 \text{ ft-lbf}}{EI}\right)(3 \text{ ft})$$
$$= 429.75 \text{ lbf-ft}^2/EI$$

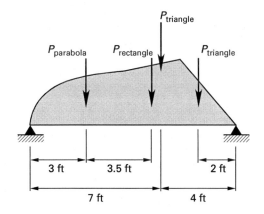

Find the conjugate reactions.

$$\sum M_R = 0 = \sum P_i x_i - R_L l$$
$$\sum M_R = P_{\text{triangle}}(2 \text{ ft}) + P_{\text{triangle}}(4 \text{ ft})$$
$$\quad + P_{\text{rectangle}}(11 \text{ ft} - 6.5 \text{ ft})$$
$$\quad + P_{\text{parabola}}(11 \text{ ft} - 3 \text{ ft}) - R_L(11 \text{ ft})$$

$$R_L(11 \text{ ft}) = \left(\frac{429.75 \text{ lbf-ft}^2}{EI}\right)(2 \text{ ft})$$
$$+ \left(\frac{20.25 \text{ lbf-ft}^2}{EI}\right)(4 \text{ ft})$$
$$+ \left(\frac{819.0 \text{ lbf-ft}^2}{EI}\right)(4.5 \text{ ft})$$
$$+ \left(\frac{910 \text{ lbf-ft}^2}{EI}\right)(8 \text{ ft})$$
$$= \frac{11{,}906 \text{ lbf-ft}^3}{EI}$$
$$R_L = 1082 \text{ lbf-ft}^2/EI$$

The conjugate moment (deflection) at 5 ft from the left support is

$$\Delta_{5\text{ft}} = M_{5\text{ft}} = R_L(5 \text{ ft}) - P_{\text{parabola}}(2 \text{ ft})$$
$$= \left(\frac{1082 \text{ lbf-ft}^2}{EI}\right)(5 \text{ ft}) - \left(\frac{910 \text{ lbf-ft}^2}{EI}\right)(2 \text{ ft})$$
$$= \frac{(3590 \text{ lbf-ft}^3)\left(12 \dfrac{\text{in}}{\text{ft}}\right)^3}{\left(1.2 \times 10^6 \dfrac{\text{lbf}}{\text{in}^2}\right)(240 \text{ in}^4)}$$
$$= 0.0215 \text{ in} \quad (2.2 \times 10^{-2} \text{ in})$$

The answer is (C).

Why Other Options Are Wrong

(A) This incorrect solution uses the moment from the loading diagram as the conjugate moment (deflection). The units do not work out.

(B) This incorrect solution does not keep track of the units in determining the conjugate moment. The equivalent loads are calculated as lbf/EI rather than lbf-ft^2/EI. When calculating the resultant deflection, the unit conversion is not correct.

(D) This solution incorrectly calculates the area of the parabola in the moment diagram. The equation for the area of a full parabola is used ($P_{\text{parabola}} = {}^4/_3 hb$) instead of that for a half-parabola.

13. The free-body diagram for the beam described is shown.

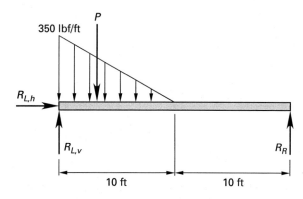

The equivalent point load for the triangular load is located at one-third the length of the load (one-third of 10 ft) and is

$$P = \tfrac{1}{2}bh = \left(\tfrac{1}{2}\right)(10 \text{ ft})\left(350 \dfrac{\text{lbf}}{\text{ft}}\right)$$
$$= 1750 \text{ lbf}$$

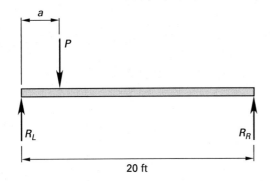

Summing moments about the left support,

$$\sum M = 0 = Pa - R_R(20 \text{ ft})$$
$$0 = (1750 \text{ lbf})\left(\frac{10 \text{ ft}}{3}\right) - R_R(20 \text{ ft})$$
$$R_R = 292 \text{ lbf}$$
$$R_L = P - R_R = 1750 \text{ lbf} - 292 \text{ lbf}$$
$$= 1458 \text{ lbf}$$

The shear diagram for this beam is as follows.

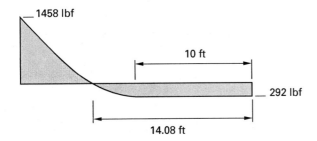

From the shear diagram, determine that the point of zero shear is between 0 and 10 ft from the left support. In this case, it is easier to write the equation for shear at a distance greater than 10 ft from the right support.

Shear is equal to the area under the load. The area under the load at a distance x from the base of the triangle is

$$A_x = \tfrac{1}{2}x\left(\frac{350 \ \tfrac{\text{lbf}}{\text{ft}}}{10 \text{ ft}}\right)x$$

The equation for shear at a distance x measured from greater than 10 ft from the right support is

$$V_x = R_R - A_x = 292 \text{ lbf} - \tfrac{1}{2}x\left(\frac{350 \ \tfrac{\text{lbf}}{\text{ft}}}{10 \text{ ft}}\right)x$$
$$= 292 \text{ lbf} - \left(\tfrac{1}{2}\right)\left(35 \ \tfrac{\text{lbf}}{\text{ft}^2}\right)x^2$$

Shear equals 0 at

$$0 = 292 \text{ lbf} - \left(17.5 \ \tfrac{\text{lbf}}{\text{ft}^2}\right)x^2$$
$$x = 4.08 \text{ ft} \quad [\text{measured left of center}]$$

The distance to the point of zero shear, measured from the left end of the beam, is

$$D_{0 \text{ shear},L} = 10 \text{ ft} - x = 10 \text{ ft} - 4.08 \text{ ft}$$
$$= 5.92 \text{ ft} \quad (5.9 \text{ ft})$$

The answer is (C).

Why Other Options Are Wrong

(A) This incorrect solution writes the shear equation at a point measured from the left support but does not include the equivalent point load in the shear equation.

The equation for shear at a distance x from the left support is

$$V_x = R_L - \left(\left(\tfrac{1}{2}\right)(10 \text{ ft} - x)\left(\frac{350 \ \tfrac{\text{lbf}}{\text{ft}}}{10 \text{ ft}}\right)(10 \text{ ft} - x)\right)$$
$$= 1458 \text{ lbf} - \left(\left(\tfrac{1}{2}\right)(10 \text{ ft} - x)\left(\frac{350 \ \tfrac{\text{lbf}}{\text{ft}}}{10 \text{ ft}}\right)(10 \text{ ft} - x)\right)$$
$$= 1458 \text{ lbf} - \left(17.5 \ \tfrac{\text{lbf}}{\text{ft}^2}\right)(10 \text{ ft} - x)^2$$

Shear equals 0 at

$$0 = 1458 \text{ lbf} - \left(17.5 \ \tfrac{\text{lbf}}{\text{ft}^2}\right)(10 \text{ ft} - x)^2$$
$$(10 \text{ ft} - x)^2 = 83.31 \text{ ft}^2$$
$$x = 0.87 \text{ ft}$$

(B) This incorrect solution calculates the distance not from the left support but rather from the center of the beam.

(D) This incorrect solution distributes the load over the entire beam instead of just one-half of it.

14. If the load from the joists acts within the center third of the wall cross section, the eccentricity from the load is negligible. For a 12 in wall, the specified dimension is 11.63 in. If the eccentricity does not exceed 1.9 in, the load acts within the center third.

$$e = \frac{11.63 \text{ in}}{2} - 2 \text{ in} = 3.82 \text{ in}$$

The eccentricity of the load must be considered.

If the load acts at 12 in on center, calculate the properties of the wall.

$$A = bh = (12 \text{ in})(11.63 \text{ in}) = 140 \text{ in}^2$$

$$S = \frac{bh^2}{6} = \frac{(12 \text{ in})(11.63 \text{ in})^2}{6} = 271 \text{ in}^3$$

The axial stress on the wall is

$$f_a = \frac{P}{A} = \frac{700 \text{ lbf}}{140 \text{ in}^2} = 5.0 \text{ lbf/in}^2$$

The bending stress on the wall creates tension on the outside face and compression on the inside face of the wall.

$$f_b = \frac{Pe}{S} = \frac{(700 \text{ lbf})(3.82 \text{ in})}{271 \text{ in}^3} = 9.87 \text{ lbf/in}^2$$

The total stress on the wall is

$$f_a + f_b = 5.0 \,\frac{\text{lbf}}{\text{in}^2} + 9.87 \,\frac{\text{lbf}}{\text{in}^2}$$

$$= 14.9 \text{ lbf/in}^2 \quad (15 \text{ lbf/in}^2 \text{ (compression)})$$

$$f_a + f_b = 5.0 \,\frac{\text{lbf}}{\text{in}^2} - 9.87 \,\frac{\text{lbf}}{\text{in}^2}$$

$$= -4.87 \text{ lbf/in}^2 \quad (5.0 \text{ lbf/in}^2 \text{ (tension)})$$

The answer is (D).

Why Other Options Are Wrong

(A) This incorrect solution neglects the effects of the eccentricity of the load (i.e., it does not calculate the bending stresses).

(B) This incorrect solution mistakenly identifies the axial stress on the wall as a tensile stress and does not consider the flexural stresses on the wall.

(C) This incorrect solution calculates the bending stresses due to the eccentricity of the load but does not add the flexural compressive stress to the axial compressive stress.

15. ACI 318 specifies that concrete exposed to freezing and thawing conditions must be air-entrained for durability (see ACI 318 Sec. 4.2). The total air content for frost-resistant concrete can be found in ACI 318 Table 4.2.1. The commentary for this section indicates that pavements, sidewalks, and parking garages are all examples of applications that experience severe exposure. For a maximum aggregate size of 1 in and severe exposure conditions, the total air content of the concrete mix should be 6.0%.

The answer is (D).

Why Other Options Are Wrong

(A) In this incorrect solution, the units on the table have been misread, and the value is "converted" to a percentage by dividing by 100.

(B) This incorrect solution uses the maximum water-cementitious materials ratio from ACI 318 Table 4.2.2 (Requirements for Special Exposure Conditions) for concrete exposed to freezing and thawing instead of the ratio from ACI 318 Table 4.2.1 (Total Air Content for Frost-Resistant Concrete).

(C) This is the value from ACI 318 Table 4.2.1 for moderate exposure, not severe exposure.

16. When slenderness must be considered in the design of compression members, the magnified moment procedure can be used if a more refined analysis is not performed.

ACI 318 Sec. 10.10.5 contains the provisions for magnified moments in nonsway frames. According to ACI 318 Sec. 10.10.1, if $kl_u/r \leq 34 - 12(M_1/M_2)$ and is no more than 40, slenderness can be ignored.

The unsupported length of a compression member is taken as the clear distance between floor slabs.

$$l_u = (13.0 \text{ ft})\left(12 \,\frac{\text{in}}{\text{ft}}\right) - 6 \text{ in} = 150 \text{ in}$$

For a 12 in diameter column,

$$r = \frac{d}{4} = \frac{12 \text{ in}}{4} = 3 \text{ in}$$

$$I_g = \frac{\pi d^4}{64} = \frac{\pi (12 \text{ in})^4}{64} = 1018 \text{ in}^4$$

$$k = 1.0 \quad [\text{ACI 318 Sec. 10.12.1}]$$

$$\frac{kl_u}{r} = \frac{(1.0)(150 \text{ in})}{3 \text{ in}} = 50$$

$$M_1 = -100 \text{ ft-kips} \quad \begin{bmatrix} \text{for columns bent in} \\ \text{double curvature} \end{bmatrix}$$

$$M_2 = 100 \text{ ft-kips}$$

$$34 - 12\left(\frac{M_1}{M_2}\right) = 34 - (12)\left(\frac{-100 \text{ ft-kips}}{100 \text{ ft-kips}}\right)$$

$$= 46 \quad [\text{so } 40 < kl_u/r < 100]$$

Therefore, slenderness must be considered, and magnified moments can be used.

The magnified moment is given by ACI 318 Eq. 10-11.

$$M_c = \delta_{\text{ns}} M_2$$

$$\delta_{ns} = \frac{C_m}{1 - \dfrac{P_u}{0.75 P_c}} \geq 1.0 \quad \text{[ACI 318 Eq. 10-12]}$$

$$P_c = \frac{\pi^2 EI}{(kl_u)^2} \quad \text{[ACI 318 Eq. 10-13]}$$

$$EI = \frac{0.4 E_c I_g}{1 + \beta_d} \quad \text{[ACI 318 Eq. 10-15]}$$

The ratio of the maximum factored axial dead load to the total factored axial load, β_{dns}, can be taken as $0.6\beta_{dns}$ according to ACI 318 Sec. R10.10.6.2.

$$EI = \frac{0.4 E_c I_g}{1 + \beta_{dns}} = \frac{(0.4)\left(3.6 \times 10^6 \, \frac{\text{lbf}}{\text{in}^2}\right)(1018 \, \text{in}^4)}{1 + 0.6}$$

$$= 9.16 \times 10^8 \, \text{lbf-in}^2$$

$$P_c = \frac{\pi^2 EI}{(kl_u)^2} = \frac{\pi^2 (9.16 \times 10^8 \, \text{lbf-in}^2)\left(\dfrac{1 \, \text{kip}}{1000 \, \text{lbf}}\right)}{((1.0)(150 \, \text{in}))^2}$$

$$= 401.8 \, \text{kips}$$

$$C_m = 0.6 + 0.4\left(\frac{M_1}{M_2}\right) \quad \text{[ACI 318 Eq. 10-16]}$$

$$= 0.6 + (0.4)\left(\frac{-100 \, \text{ft-kips}}{100 \, \text{ft-kips}}\right)$$

$$= 0.2 \quad [\leq 0.4. \text{ Use } 0.4.]$$

$$\delta_{ns} = \frac{C_m}{1 - \dfrac{P_u}{0.75 P_c}} = \frac{0.4}{1 - \dfrac{300 \, \text{kips}}{(0.75)(401.8 \, \text{kips})}}$$

$$= 89.3 \quad [\geq 1.0, \text{ OK}]$$

$$M_c = \delta_{ns} M_2 = (89.3)(100 \, \text{ft-kips})$$

$$= 8930 \, \text{ft-kips} \quad (9000 \, \text{ft-kips})$$

It is interesting to note that even for kl_u/r just over 40 (50 in this case), the moment magnifier, δ_{ns}, is large.

The answer is (C).

Why Other Options Are Wrong

(A) This incorrect solution uses the wrong values for EI in ACI 318 Eq. 10-13 for the critical load, P_c. EI as used in ACI 318 Eq. 10-13 is given by Eq. 10-14 or Eq. 10-15. This wrong solution uses $E_c I$.

(B) This incorrect solution ignores the lower limit for C_m in ACI 318 Eq. 10-16.

(D) This incorrect solution uses a positive sign for the M_1/M_2 ratio. A positive sign indicates single curvature, not double.

17. The tributary width for an interior column is found by summing the halved distances to the adjacent column(s) from the column centerline.

$$w' = \sum \tfrac{1}{2} d$$

$$w'_1 = \left(\tfrac{1}{2}\right)(20 \, \text{ft}) + \left(\tfrac{1}{2}\right)(20 \, \text{ft}) = 20 \, \text{ft}$$

$$w'_2 = \left(\tfrac{1}{2}\right)(20 \, \text{ft}) + \left(\tfrac{1}{2}\right)(20 \, \text{ft}) = 20 \, \text{ft}$$

The tributary area for an interior column is

$$A = w'_1 w'_2 = (20 \, \text{ft})(20 \, \text{ft})$$

$$= 400 \, \text{ft}^2$$

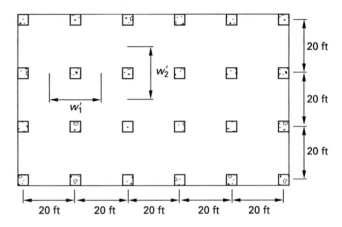

From IBC Sec. 1605.3, the applicable service load combinations are

$$D + L_{\text{floor}}$$

$$D + L_{\text{roof}}$$

$$D + 0.75(L_{\text{floor}} + L_{\text{roof}})$$

The total uniform load per unit area is the larger of

$$w_{\text{uniform}} = (D_{\text{roof}} + D_{\text{floor}}) + L_{\text{floor}}$$

$$= \left(15 \, \frac{\text{lbf}}{\text{ft}^2} + 15 \, \frac{\text{lbf}}{\text{ft}^2}\right) + 60 \, \frac{\text{lbf}}{\text{ft}^2}$$

$$= 90 \, \text{lbf/ft}^2$$

$$w_{\text{uniform}} = (D_{\text{roof}} + D_{\text{floor}}) + L_{\text{roof}}$$

$$= \left(15 \, \frac{\text{lbf}}{\text{ft}^2} + 15 \, \frac{\text{lbf}}{\text{ft}^2}\right) + 20 \, \frac{\text{lbf}}{\text{ft}^2}$$

$$= 50 \, \text{lbf/ft}^2$$

$$w_{\text{uniform}} = (D_{\text{roof}} + D_{\text{floor}}) + 0.75(L_{\text{floor}} + L_{\text{roof}})$$

$$= \left(15 \, \frac{\text{lbf}}{\text{ft}^2} + 15 \, \frac{\text{lbf}}{\text{ft}^2}\right) + (0.75)\left(60 \, \frac{\text{lbf}}{\text{ft}^2} + 20 \, \frac{\text{lbf}}{\text{ft}^2}\right)$$

$$= 90 \, \text{lbf/ft}^2$$

The total column load is

$$P = w_{\text{uniform}} A = \left(90 \ \frac{\text{lbf}}{\text{ft}^2}\right)(400 \ \text{ft}^2)$$
$$= 36{,}000 \ \text{lbf} \quad (36 \ \text{kips})$$

The answer is (C).

Why Other Options Are Wrong

(A) This incorrect solution uses tributary width instead of tributary area when calculating the total load on the column.

(B) This incorrect solution does not include the second-floor loads when calculating the total load on the column.

(D) This incorrect solution does not consider the applicable load combinations and instead uses the total roof and floor live and dead loads.

18. Section 1607.11 of the IBC states that roofs must be designed for the appropriate live loads. The minimum uniformly distributed roof live loads are given in IBC Table 1607.1 and are permitted to be reduced per IBC Eq. 16-25. The minimum roof live load, L_o, is given in IBC Table 1607.1 as 20 lbf/ft^2 for an ordinary pitched roof. The reduced live load is

$$L_r = L_o R_1 R_2 \quad \text{[IBC Eq. 16-25]}$$

The reduction factors are based on the tributary area of the structural member and the slope of the roof. The tributary area of the column is

$$A_t = (20 \ \text{ft})(15 \ \text{ft}) = 300 \ \text{ft}^2$$

The number of inches of rise per foot of the roof is

$$F = 6$$

The reduction factors are calculated using IBC Eq. 16-27 for R_1 if 200 ft$^2 < A_t <$ 600 ft^2 and using IBC Eq. 16-30 for R_2 if $4 < F < 12$.

$$R_1 = 1.2 - 0.001 A_t = 1.2 - (0.001)(300)$$
$$= 0.90$$
$$R_2 = 1.2 - 0.05 F = 1.2 - (0.05)(6)$$
$$= 0.90$$

The reduced live load is

$$L_r = L_o R_1 R_2 = \left(20 \ \frac{\text{lbf}}{\text{ft}^2}\right)(0.90)(0.90)$$
$$= 16.2 \ \text{lbf/ft}^2$$

The minimum live load on the column is

$$P = L_r A_t = \left(16.2 \ \frac{\text{lbf}}{\text{ft}^2}\right)(300 \ \text{ft}^2)\left(\frac{1 \ \text{kip}}{1000 \ \text{lbf}}\right)$$
$$= 4.86 \ \text{kips} \quad (4.9 \ \text{kips})$$

The answer is (B).

Why Other Options Are Wrong

(A) This incorrect solution uses the live load reduction factors found in IBC Sec. 1607.9 instead of the roof reduction factors.

(C) This incorrect solution does not reduce the roof load.

(D) This incorrect solution calculates the total load, not the live load.

19. Using the moment-area method, the deflection of a beam at a particular point is equal to the moment of the M/EI diagram about that point.

The moment at the fixed end of the beam shown is

$$M = Pl = (10 \ \text{kips})(25 \ \text{ft}) = 250 \ \text{ft-kips}$$

The moment diagram is

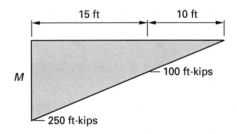

To draw the M/EI diagram, divide the moment diagram by the respective moment of inertia.

$$\frac{M_{\text{support}}}{EI} = \frac{250 \ \text{ft-kips}}{E(2000 \ \text{in}^4)} = \frac{0.125 \ \text{ft-kips}}{E \ \text{in}^4}$$

The M/EI diagram is

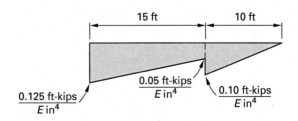

The deflection at the free end of the beam is the moment of the M/EI diagram about the free end.

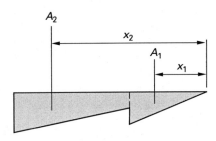

$$\delta = A_1 x_1 + A_2 x_2$$
$$= \left(\tfrac{1}{2}\right)(10\ \text{ft})\left(\frac{0.10\ \text{ft-kips}}{E\ \text{in}^4}\right)\left(\tfrac{2}{3}\right)(10\ \text{ft})$$
$$+ \left(\frac{0.05\ \text{ft-kips}}{E\ \text{in}^4}\right)(15\ \text{ft})\left(10\ \text{ft} + \frac{15\ \text{ft}}{2}\right)$$
$$+ \left(\tfrac{1}{2}\right)(15\ \text{ft})\left(\frac{0.125\ \text{ft-kips} - 0.05\ \text{ft-kips}}{E\ \text{in}^4}\right)$$
$$\times \left((15\ \text{ft})\left(\tfrac{2}{3}\right) + 10\ \text{ft}\right)$$
$$= \frac{27.7\ \frac{\text{ft}^3\text{-kips}}{\text{in}^4}}{E}$$
$$= \frac{\left(27.7\ \frac{\text{ft}^3\text{-kips}}{\text{in}^4}\right)\left(12\ \frac{\text{in}}{\text{ft}}\right)^3\left(1000\ \frac{\text{lbf}}{\text{kip}}\right)}{29 \times 10^6\ \frac{\text{lbf}}{\text{in}^2}}$$
$$= 1.65\ \text{in}\quad (1.7\ \text{in})$$

The answer is (B).

Why Other Options Are Wrong

(A) This incorrect solution does not convert kips to pounds in calculating the deflection. The units do not work out.

(C) This incorrect solution reverses I_1 and I_2 when drawing the M/EI diagram.

(D) This incorrect solution does not make the adjustment for the differing moment of inertia over the span length. The deflection is incorrectly calculated using a moment of inertia of 1000 in^4 for the entire length of the beam.

20. A roof member that does not support a ceiling is limited by Table 1604.3 of the IBC to a total load deflection of $l/120$.

$$\Delta = \frac{l}{120} = \left(\frac{60\ \text{ft}}{120}\right)\left(12\ \frac{\text{in}}{\text{ft}}\right) = 6.0\ \text{in}\quad (6\ \text{in})$$

The answer is (D).

Why Other Options Are Wrong

(A) This incorrect solution does not convert the length of the member to inches in calculating the deflection limit. The units do not work out.

(B) This incorrect solution uses the deflection limit for total load with plastered ceilings, $l/240$, instead of the deflection limit for total load with no ceiling.

(C) This incorrect solution uses the total load deflection limit for nonplastered ceilings, $l/180$, instead of the deflection limit for no ceiling.

21. Frequent variations or reversals of stress can cause fatigue. Appendix 3 of the AISC *Steel Construction Manual* gives requirements for fatigue loading. If the number of loading cycles exceeds 20,000 over the lifetime of the member, fatigue must be considered.

Determine the number of cycles over the lifetime of this structure.

$$N = \left(\frac{\text{cycles}}{\text{day}}\right)(\text{number of years})\left(365\ \frac{\text{days}}{\text{yr}}\right)$$
$$= \left(200\ \frac{\text{cycles}}{\text{day}}\right)(25\ \text{years})\left(365\ \frac{\text{days}}{\text{yr}}\right)$$
$$= 1{,}825{,}000\ \text{cycles}\quad \left[\begin{array}{l}> 20{,}000\ \text{cycles.}\\ \text{Fatigue has to be considered.}\end{array}\right]$$

Two questions are used to determine whether App. 3 is applicable for use in fatigue design.

1. Is the member subject to normal atmospheric conditions?
2. Is it subject to temperatures less than 300°F?

For this problem, the answer to both questions is "yes." Therefore, App. 3 is applicable, and the maximum stress for unfactored loads shall be $\leq 0.66 F_y$. Note that service loads, not factored loads, are used for fatigue design.

From AISC *Steel Construction Manual* Table A-3.1, determine that a bolted end connection is shown in Ex. 2.1 and Ex. 2.2. From AISC Table A-3.1, determine that

stress category   B
constant, C_f 120×10^8
threshold, F_{TH} 16 kips/in^2

The design stress range, F_{sr} (in kips/in^2), is given as

$$F_{sr} = \left(\frac{C_f}{N}\right)^{0.333}\quad [\text{AISC Eq. A-3.1}]$$
$$= \left(\frac{120 \times 10^8}{1{,}825{,}000}\right)^{0.333}$$
$$= 18.68\ \text{kips/in}^2\quad [> F_{TH}]$$

Use F_{sr} equal to 18.68 kips/in².

Determine the range of loads.

$$P_{max} = D + L = 20 \text{ kips} + 50 \text{ kips}$$
$$= 70 \text{ kips (tension)}$$
$$P_{min} = D - L = 20 \text{ kips} - 10 \text{ kips}$$
$$= 10 \text{ kips (tension)}$$

The allowable tensile stress is

$$F_t = 0.66F_y = (0.66)\left(36 \frac{\text{kips}}{\text{in}^2}\right) = 23.8 \text{ kips/in}^2$$

Estimate the channel size.

$$A_{req} = \frac{P_{max}}{F_t} = \frac{70 \text{ kips}}{23.8 \frac{\text{kips}}{\text{in}^2}} = 2.94 \text{ in}^2$$

Try a C8 × 11.5.

$$A_{prov} = 3.38 \text{ in}^2$$
$$f_{t,max} = \frac{P_{max}}{A_{prov}} = \frac{70 \text{ kips}}{3.38 \text{ in}^2} = 20.7 \text{ kips/in}^2$$
$$f_{t,min} = \frac{P_{min}}{A_{prov}} = \frac{10 \text{ kips}}{3.38 \text{ in}^2} = 2.96 \text{ kips/in}^2$$

The actual stress range is

$$f_{sr} = 20.7 \frac{\text{kips}}{\text{in}^2} - 2.96 \frac{\text{kips}}{\text{in}^2} = 17.7 \text{ kips/in}^2$$

The allowable (design) stress range is

$$F_{sr} = 18.68 \text{ kips/in}^2 \quad [> 17.7 \text{ kips/in}^2, \text{ OK}]$$

Use a C8 × 11.5.

The answer is (C).

Why Other Options Are Wrong

(A) This incorrect solution adds the alternating live loads together instead of treating them as separate loading conditions.

(B) This incorrect solution uses a design stress range of 16 kips/in², which is the threshold stress.

(D) This incorrect solution uses the threshold stress range as the actual design stress.

22. Section I3 of the AISC *Steel Construction Manual* states that the total horizontal shear to be resisted between the point of maximum positive moment and the points of zero moment is the lesser of

$$V' = 0.85 f'_c A_c = (0.85)\left(3 \frac{\text{kips}}{\text{in}^2}\right)(288 \text{ in}^2)$$
$$= 734 \text{ kips} \quad [\text{AISC Eq. I3-1a}]$$
$$V' = F_y A_s = \left(50 \frac{\text{kips}}{\text{in}^2}\right)(11.8 \text{ in}^2)$$
$$= 590 \text{ kips} \quad [\text{AISC Eq. I3-1b}]$$

The horizontal shear force that must be carried by the studs is 590 kips.

$$V' = \sum Q_n \quad [\text{AISC Eq. I3-1c}]$$
$$590 \text{ kips} = \sum Q_n$$

The allowable horizontal shear load on a ³⁄₄ in × 3¹⁄₂ in headed stud in the strong direction and lightweight concrete is found from AISC Table 3-21.

$$Q_n = 17.1 \text{ kips/stud}$$

The number of studs required between the point of maximum positive moment and the points of zero moment is

$$n = \frac{\sum Q_n}{Q_n} = \frac{590 \text{ kips}}{17.1 \frac{\text{kips}}{\text{stud}}}$$
$$= 34.5 \text{ studs} \quad [\text{Use 35 studs.}]$$

For a simply supported beam, the maximum positive moment occurs at the midspan of the beam. Therefore, the total number of shear studs required is

$$n_{total} = 2n = (2)(35 \text{ studs}) = 70 \text{ studs}$$

The answer is (C).

Why Other Options Are Wrong

(A) This incorrect solution only calculates the number of studs required for one-half of the span.

(B) This incorrect solution uses the value for normal-weight concrete in determining the stud capacity.

(D) This incorrect solution uses the concrete shear as the limiting V' instead of the lesser value.

23. Determine the design loads.

$$P_u = 1.2D + 1.6L$$
$$= (1.2)(130 \text{ kips}) + (1.6)(390 \text{ kips})$$
$$= 780 \text{ kips}$$
$$P_u \leq \phi_c P_n$$
$$780 \text{ kips} \leq \phi_c P_n$$

The values in the columns tables in the AISC *Steel Construction Manual* are based on an effective length with respect to the minor axis, KL_y.

The effective column lengths are given as

$$KL_x = 32 \text{ ft}$$
$$KL_y = 18 \text{ ft}$$

Enter the AISC columns tables with an effective length of 18 ft. Since the column depth is limited to 12 in, begin with the W12 sections. For $\phi_c P_n$, use the unshaded columns and determine that a W12 × 87 column has a capacity of 801 kips. Check the capacity based on the effective length for buckling about the x-axis.

From the AISC column tables, $r_x/r_y = 1.75$.

The equivalent effective length for the x-axis is

$$\frac{L_x}{\frac{r_x}{r_y}} = \frac{32 \text{ ft}}{1.75} = 18.3 \text{ ft} \quad [>18 \text{ ft}]$$

Therefore, check the capacity for $KL = 18.3$ ft.

By interpolation, a W12 × 87 column has a capacity of 792 kips [OK].

The answer is (A).

Why Other Options Are Wrong

(B) This solution incorrectly uses the column for ASD instead of LRFD in the column tables to find that a W12 × 136 column is needed.

(C) This incorrect solution reverses the major and minor axes in determining effective length.

(D) This incorrect solution reverses the major and minor axes in determining effective length and uses the column for ASD instead of LRFD in the column tables.

24. Use the column base plates design procedure given in Part 14 of the AISC *Steel Construction Manual*.

For a W12 × 72 steel column,

$$d = 12.3 \text{ in}$$
$$b_f = 12.0 \text{ in}$$

Determine the required design loads.

$$P_u = 1.2D + 1.6L = (1.2)(105 \text{ kips}) + (1.6)(315 \text{ kips})$$
$$= 630 \text{ kips}$$

Determine the bearing stress.

$$f_u = \frac{P_u}{NB} = \frac{630 \text{ kips}}{(16 \text{ in})(14 \text{ in})} = 2.81 \text{ kips/in}^2$$

Determine the available bearing strength, where ϕ_c is 0.60.

$$\phi_c P_p = \phi_c 0.85 f'_c A_1 \sqrt{\frac{A_2}{A_1}}$$

$$A_1 = NB = (16 \text{ in})(14 \text{ in}) = 224 \text{ in}^2$$

$$A_2 = L_{\text{ftg}} W_{\text{ftg}} = \left((8 \text{ ft})\left(12 \frac{\text{in}}{\text{ft}}\right)\right)\left((8 \text{ ft})\left(12 \frac{\text{in}}{\text{ft}}\right)\right)$$
$$= 9216 \text{ in}^2$$

$$\sqrt{\frac{A_2}{A_1}} = \sqrt{\frac{9216 \text{ in}^2}{224 \text{ in}^2}}$$
$$= 6.41 \quad [>2.0. \text{ Use } 2.0.]$$

$$\phi_c P_p = \phi_c 0.85 f'_c A_1 \sqrt{\frac{A_2}{A_1}}$$
$$= (0.60)(0.85)\left(3 \frac{\text{kips}}{\text{in}^2}\right)(224 \text{ in}^2)(2.0)$$
$$= 685.44 \text{ kips}$$

The thickness of the base plate is

$$t_p = l\sqrt{\frac{2f_u}{0.9 F_y}}$$

Determine the critical cantilever projection of the base plate, l. Use the larger of m, n, and $\lambda n'$.

$$m = \frac{N - 0.95d}{2} = \frac{16 \text{ in} - (0.95)(12.3 \text{ in})}{2}$$
$$= 2.16 \text{ in}$$

$$n = \frac{B - 0.80 b_f}{2} = \frac{14 \text{ in} - (0.80)(12.0 \text{ in})}{2}$$
$$= 2.20 \text{ in}$$

$$n' = \frac{\sqrt{db_f}}{4} = \frac{\sqrt{(12.3 \text{ in})(12.0 \text{ in})}}{4}$$
$$= 3.04 \text{ in}$$

Determine λ.

$$\lambda = \frac{2\sqrt{X}}{1 + \sqrt{1 - X}} \leq 1.0$$

$$X = \frac{4 db_f P_u}{(d + b_f)^2 (\phi_c P_p)}$$
$$= \frac{(4)(12.3 \text{ in})(12.0 \text{ in})(630 \text{ kips})}{(12.3 \text{ in} + 12.0 \text{ in})^2 (685.44 \text{ kips})}$$
$$= 0.92$$

$$\lambda = \frac{2\sqrt{X}}{1+\sqrt{1-X}} = \frac{2\sqrt{0.92}}{1+\sqrt{1-0.92}}$$
$$= 1.50 \quad [>1.0. \text{ Use } 1.0.]$$
$$\lambda n' = (1.0)(3.04 \text{ in})$$
$$= 3.04 \text{ in} \quad [\text{governs}]$$

Determine the required base plate thickness, where $l = \lambda n'$.

$$t_{req} = l\sqrt{\frac{2f_u}{0.9F_y}} = 3.04 \text{ in}\sqrt{\frac{(2)\left(2.81 \frac{\text{kips}}{\text{in}^2}\right)}{(0.9)\left(36 \frac{\text{kips}}{\text{in}^2}\right)}}$$
$$= 1.27 \text{ in}$$

The answer is (C).

Why Other Options Are Wrong

(A) This incorrect solution neglects to calculate $\lambda n'$, which in this case is the controlling distance.

(B) This incorrect solution uses F_u instead of F_y when calculating the required thickness.

(D) This incorrect solution reverses N and B when calculating m and n. N is the length of the base plate in the direction of the column depth. B is the width of the base plate in the direction of the column flange, as shown in Fig. 14-3 of the AISC column base plates design procedure.

25. Determine the eccentricity of the load. If the resultant falls outside the column flanges, the anchor bolts must resist the resulting tension.

$$e = \frac{M}{P} = \frac{(200 \text{ ft-kips})\left(12 \frac{\text{in}}{\text{ft}}\right)}{320 \text{ kips}} = 7.5 \text{ in}$$

For a W14 × 109 column,

$$d = 14.32 \text{ in}$$
$$\frac{d}{2} = \frac{14.32 \text{ in}}{2} = 7.16 \text{ in}$$
$$t_f = 0.860 \text{ in}$$
$$\frac{t_f}{2} = \frac{0.860 \text{ in}}{2} = 0.430 \text{ in}$$
$$\frac{d}{2} - \frac{t_f}{2} = 7.16 \text{ in} - 0.43 \text{ in} = 6.73 \text{ in}$$

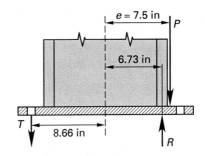

The eccentricity is outside the column flange. Therefore, the bolt must resist uplift. Assume the resultant of the compression forces is located at the center of the column flange. Take moments about this point.

$$\sum M_R = 0 = P\left(e - \left(\frac{d}{2} - \frac{t_f}{2}\right)\right)$$
$$- T\left(\left(\frac{d}{2} + 1.5 \text{ in}\right) + \left(\frac{d}{2} - \frac{t_f}{2}\right)\right)$$
$$0 = (320 \text{ kips})(7.5 \text{ in} - 6.73 \text{ in})$$
$$- T(8.66 \text{ in} + 6.73 \text{ in})$$
$$T = 16.0 \text{ kips}$$

Calculate the size of bolt required. From Table 7-2 of the AISC *Steel Construction Manual*, determine that a 1 in diameter bolt has an allowable tensile load of 17.7 kips.

The answer is (D).

Why Other Options Are Wrong

(A) This incorrect solution does not consider the moment in the design. For a base plate without any uplift, a ⅝ in diameter bolt is adequate.

(B) This incorrect solution uses the distance to the edge of the column flange instead of the distance to the center of the flange when calculating the tension on the bolt.

(C) This incorrect solution uses the LRFD instead of the ASD column value in the bolt Available Tensile Strength table.

26. The weight of brick veneer is given as 40 lbf/ft². Brick veneer is typically supported at each floor. Therefore, the brick load on spandrel beam BC is for a one-story height, or 13 ft. The spacing between beams is 10 ft. The tributary width, w', for a spandrel beam is

$$w' = \left(\tfrac{1}{2}\right)(10 \text{ ft}) = 5 \text{ ft}$$

The load on the spandrel beam is

$$w_{\text{per ft}} = w' w_{\text{per ft}^2}$$

$$w_D = (5 \text{ ft})\left(60 \frac{\text{lbf}}{\text{ft}^2}\right) = 300 \text{ lbf/ft}$$

$$w_L = (5 \text{ ft})\left(40 \frac{\text{lbf}}{\text{ft}^2}\right) = 200 \text{ lbf/ft}$$

$$w_{\text{brick}} = (13 \text{ ft})\left(40 \frac{\text{lbf}}{\text{ft}^2}\right) = 520 \text{ lbf/ft}$$

$$\begin{aligned} w_{\text{total}} &= w_D + w_L + w_{\text{brick}} \\ &= 300 \frac{\text{lbf}}{\text{ft}} + 200 \frac{\text{lbf}}{\text{ft}} + 520 \frac{\text{lbf}}{\text{ft}} \\ &= 1020 \text{ lbf/ft} \end{aligned}$$

The maximum moment on the beam is

$$M_{\max} = \frac{w_{\text{total}} L^2}{8} = \frac{\left(1020 \frac{\text{lbf}}{\text{ft}}\right)(30 \text{ ft})^2}{8}$$
$$= 114{,}750 \text{ ft-lbf} \quad (115 \text{ ft-kips})$$

Using the AISC *Steel Construction Manual*'s Selection Table, the lightest beam required based on maximum moment is a W14 × 30 (in boldface type) ($M_n/\Omega = 118$ ft-kips).

Check for deflection. The maximum deflection of the beam cannot exceed $L/600$ or 0.3 in, whichever is less.

$$\frac{L}{600} = \frac{(30 \text{ ft})\left(12 \frac{\text{in}}{\text{ft}}\right)}{600} = 0.6 \text{ in}$$

Therefore, the deflection cannot exceed 0.3 in.

Deflection for a simply supported beam is

$$\Delta = \frac{5 w_{\text{total}} L^4}{384 EI} \leq 0.3 \text{ in}$$

Solving for I gives

$$\begin{aligned} I &\geq \frac{5 w_{\text{total}} L^4}{384 E \Delta} \\ &\geq \frac{(5)\left(1020 \frac{\text{lbf}}{\text{ft}}\right)(30 \text{ ft})^4 \left(12 \frac{\text{in}}{\text{ft}}\right)^3}{(384)\left(29 \times 10^6 \frac{\text{lbf}}{\text{in}^2}\right)(0.3 \text{ in})} \\ &\geq 2137 \text{ in}^4 \end{aligned}$$

Using the AISC *Steel Construction Manual*'s Moment of Inertia Selection Table, the lightest beam required based on deflection is a W24 × 84 (in boldface type) ($I_x = 2370 \text{ in}^4$). Deflection controls. Use a W24 × 84 beam.

Check the additional deflection due to the self-weight of the beam.

$$\begin{aligned} \Delta &= \frac{5 w_{\text{total}} L^4}{384 EI} \\ &= \frac{(5)\left(84 \frac{\text{lbf}}{\text{ft}} + 1020 \frac{\text{lbf}}{\text{ft}}\right)(30 \text{ ft})^4 \left(12 \frac{\text{in}}{\text{ft}}\right)^3}{(384)\left(29 \times 10^6 \frac{\text{lbf}}{\text{in}^2}\right)(2370 \text{ in}^4)} \\ &= 0.29 \text{ in} \quad [\leq 0.3 \text{ in, OK}] \end{aligned}$$

For relatively short spans (generally 15 ft or less for all but the largest sections), shear stress in the web should also be checked. In this case the span is 30 ft, so bending stress controls.

The answer is (D).

Why Other Options Are Wrong

(A) This incorrect solution neglects the weight of the facade when calculating the load on the spandrel beam and ignores the deflection limitation of $L/600$ or 0.3 in.

(B) This incorrect solution correctly calculates the moment on the beam but ignores the deflection limitation of $L/600$ or 0.3 in.

(C) This incorrect solution correctly calculates the moment on the beam but uses the wrong deflection limitation. The deflection is limited to $L/600$ or 0.3 in, whichever is less. In this case, $L/600$ is greater than 0.3 in, so the deflection should be limited to 0.3 in.

27. The equivalent axial load procedure can be used to design beam-columns.

$$P_e = P + M_x m + M_y m U$$
$$M_x = P e_x = (400 \text{ kips})(12 \text{ in})\left(\frac{1 \text{ ft}}{12 \text{ in}}\right) = 400 \text{ ft-kips}$$
$$M_y = 0$$

Try a W14 shape.

$$m = \frac{24 \text{ in}}{d} = \frac{24 \text{ in}}{14 \text{ in-ft}} = 1.71 \text{ ft}^{-1}$$
$$\begin{aligned} P_e &= 400 \text{ kips} + (400 \text{ ft-kips})\left(1.71 \frac{1}{\text{ft}}\right) + 0 \text{ kip} \\ &= 1084 \text{ kips} \end{aligned}$$

Estimate a column size using AISC Axial Compression Table 4-1. For an unbraced length of 25 ft using ASD, try a W14 × 193 section. By interpolation, a W14 × 193 section has a capacity of 1135 kips at an unbraced length of 25 ft. Check if the section works using AISC *Specification for Structural Steel Buildings* (AISC *Specification*) Sec. H1.1.

If $P_r/P_c \geq 0.2$, then use the modified form of AISC *Specification* Eq. H1-1a, as given in Part 6 of the AISC *Manual*.

$$\frac{P_r}{P_c} = \frac{P_r}{\frac{P_n}{\Omega_c}} = \frac{400 \text{ kips}}{1135 \text{ kips}} = 0.35 \quad [\geq 0.2]$$

Use AISC Eq. H1-1a.

$$pP_r + b_x M_{rx} + b_y M_{ry} \leq 1.0 \quad [\text{AISC Eq. H1-1a}]$$

From the Combined Axial and Bending Table 6-1, for a $W14 \times 193$ with an effective length of 25 ft,

$$p = 0.880 \times 10^{-3} \text{ kip}^{-1}$$
$$b_x = 1.07 \times 10^{-3} \text{ (ft-kip)}^{-1}$$

$$pP_r + b_x M_{rx} + b_y M_{ry} = \left(0.880 \times 10^{-3} \frac{1}{\text{kip}}\right)(400 \text{ kips})$$
$$+ \left(\left(1.07 \times 10^{-3} \frac{1}{\text{ft-kip}}\right) \times (400 \text{ ft-kips})\right)$$
$$+ 0$$
$$= 0.78 \quad [\leq 1.0]$$

A $W14 \times 193$ will work, but it is conservative. Try a smaller section.

Check a $W14 \times 159$. By interpolation, for an unbraced length of 25 ft,

$$\frac{P_n}{\Omega_c} = 927.5 \text{ kips}$$
$$\frac{P_r}{P_c} = \frac{P_r}{\frac{P_n}{\Omega_c}} = \frac{400 \text{ kips}}{927.5 \text{ kips}} = 0.43 \quad [> 0.2]$$

Use Eq. H1-1a.

From Table 6-1, for a $W14 \times 159$ and an effective length of 25 ft,

$$p = 1.075 \times 10^{-3} \text{ kip}^{-1}$$
$$b_x = 1.35 \times 10^{-3} \text{ (ft-kip)}^{-1}$$
$$pP_r + b_x M_{rx} + b_y M_{ry} = \left(1.075 \times 10^{-3} \frac{1}{\text{kip}}\right)(400 \text{ kips})$$
$$+ \left(\left(1.35 \times 10^{-3} \frac{1}{\text{ft-kip}}\right) \times (400 \text{ ft-kips})\right)$$
$$+ 0$$
$$= 0.97 \quad [\leq 1.0]$$

Use a $W14 \times 159$.

The answer is (B).

Why Other Options Are Wrong

(A) This incorrect solution mixes the LRFD coefficients in Table 6-1 with the ASD equations.

(C) This incorrect solution does not perform the necessary number of iterations. The procedure should be repeated until the unity equation approaches 1.0.

(D) This incorrect solution assumes the eccentricity about the wrong axis.

28. The nominal shear strength of a fillet weld is given in Table J2.5 of the AISC *Steel Construction Manual* as

$$F_w = 0.60 F_{\text{EXX}}$$

The nominal strength of the weld metal is determined by the electrodes used. E70XX electrodes have a strength of 70 kips/in^2.

The available weld length can be taken as the workable flat dimension or the nominal dimension of the tube less the radius at each end. According to the AISC *Specification*, the outside corner radii are taken as $2.25 t_{\text{nom}}$.

$$L_w = 2(b - (2)(2.25 t_{\text{nom}}))$$
$$= (2)(4 \text{ in} - (2)(2.25)(0.375 \text{ in}))$$
$$= 4.63 \text{ in}$$

The effective throat, t_e, of a fillet weld using the SMAW process is $0.707w$. The shear capacity of a weld is given by AISC Eq. J2-3 as

$$R_n = F_w A_w = 0.60 F_{\text{EXX}} L_w t_e$$
$$= (0.60)\left(70 \frac{\text{kips}}{\text{in}^2}\right)(4.63 \text{ in}) t_e$$

The allowable strength, R_n/Ω, must equal or exceed the load on the weld.

$$12{,}000 \text{ lbf} = \frac{R_n}{2.00}$$

Solving for the required effective throat thickness,

$$t_e = \frac{(12{,}000 \text{ lbf})(2.00)}{\left(42 \dfrac{\text{kips}}{\text{in}^2}\right)(4.63 \text{ in})\left(1000 \dfrac{\text{lbf}}{\text{kip}}\right)}$$
$$= 0.123 \text{ in}$$

The nominal weld size is

$$w = \frac{t_e}{0.707} = \frac{0.123 \text{ in}}{0.707}$$
$$= 0.174 \text{ in}$$

A $^3/_{16}$ in fillet weld provides adequate capacity.

Check the minimum weld size required based on the thicknesses of the materials. AISC Table J2.4 specifies the minimum weld size based on the thickness of the thinner part joined.

The thickness of the W10 × 33 flange is found in AISC Part 1.

$$t_f = 0.435 \text{ in}$$

The thickness of the HSS 4 × 4 × $^3/_8$ tube is $^3/_8$ in or 0.375 in. Therefore, the tube thickness controls. The minimum weld size from AISC Table J2.4 is $^3/_{16}$ in.

Use a $^3/_{16}$ in fillet weld.

The answer is (B).

Why Other Options Are Wrong

(A) This incorrect solution uses the nominal tube length in determining the weld length and does not check the minimum weld size based on the thicknesses of the materials.

(C) This incorrect solution uses the flange thickness of a W10 × 39 column instead of a W10 × 33 column. In this case, the thicknesses of materials controls the size of the weld.

(D) This incorrect solution does not use the same units for the nominal weld strength and the load on the weld when calculating the required weld size.

29. Part 10 of the AISC *Steel Construction Manual* covers connection design. Using the table for double angle connections, the maximum beam reaction is the lesser of the available strength of the bolt and angle and the available strength of the beam web.

$$L3\tfrac{1}{2} \times 3\tfrac{1}{2} \times \tfrac{5}{16} \text{ angles}$$

$$\tfrac{3}{4} \text{ in diameter A325} - \text{N bolts}$$

$$L_{eh} = 1.75 \text{ in}$$

$$L_{ev} = 1.5 \text{ in}$$

$$t_w = 0.23 \text{ in}$$

From AISC Table 10-1, determine the available strength of the bolt and angle to be 95.4 kips.

The beam web available strength is 207 kips/in of beam web thickness.

$$R_u = t_w(\text{tabulated value}) = (0.23 \text{ in})\left(207 \, \frac{\text{kips}}{\text{in}}\right)$$
$$= 47.6 \text{ kips}$$

The beam web available strength controls. The maximum reaction is 47.6 kips (48 kips).

The answer is (C).

Why Other Options Are Wrong

(A) This incorrect solution uses the ASD column in the table for beam web available strength.

(B) This incorrect solution mistakenly uses the value for L_{eh} equal to 1.5 in, instead of 1.75 in, to determine the beam web available strength.

(D) This incorrect solution only considers the bolt and angle available strength and does not check the beam web available strength.

30. The welds in this case are subject to eccentric loading. The minimum weld size, in sixteenths of an inch, can be determined using ASD from the equation

$$D_{\min} = \frac{\Omega P_a}{CC_1 l}$$

From Table 8-4 of the AISC *Steel Construction Manual*,

$$\Omega = 2.00$$

$$C_1 = 1.0 \text{ for E70XX electrodes}$$

$$k = 0 \text{ for special case of load not in the plane of the weld group}$$

The characteristic length of the weld group, l, is given as 8 in, and the eccentricity is given as 2.25 in. To determine the coefficient, C, in AISC Table 8-4, first determine

$$a = \frac{e_x}{l} = \frac{2.25 \text{ in}}{8 \text{ in}} = 0.28$$

By interpolation, C is equal to 3.18 kips/in. The minimum weld size is

$$D_{\min} = \frac{\Omega P_a}{CC_1 l} = \frac{(2.00)(40 \text{ kips})}{\left(3.18 \, \dfrac{\text{kips}}{\text{in}}\right)(1.0)(8 \text{ in})} = 3.14$$

The minimum thickness of the weld is

$$t_w = \frac{D_{\min}}{16} = \frac{3.14}{16} \text{ in}$$
$$= 4/16 \text{ in} \quad (1/4 \text{ in})$$

Check the minimum weld size based on the thickness of the parts joined, using AISC Table J2.4. The flange thickness of a W12 × 50 is $^5/_8$ in, and the thickness of the angle is $^1/_2$ in. The minimum weld size based on the thinner part joined is $^3/_{16}$ in, which is less than the $^1/_4$ in required. Use a $^1/_4$ in weld.

The answer is (C).

Why Other Options Are Wrong

(A) This incorrect solution uses a weld length of 16 in rather than 8 in.

(B) The weld size is based on shear alone. The eccentricity of the load is ignored in this incorrect solution.

(D) This incorrect solution applies the ϕ factor to the ASD equation for weld thickness.

31. Section 4 of the AISC *Steel Construction Manual* covers the design of compression members. The Available Strength in Axial Compression, AISC Table 4-9, can be used to find the critical compressive stress for the double angle in this case.

For an L8 × 6 × ½ double angle with Kl_x of 23 ft, by interpolation of Table 4-9,

$$\frac{P_n}{\Omega_c} = 151.5 \text{ kips}$$

The critical compressive stress is

$$F_{\text{cr}} = \frac{P_n}{A_g} = \frac{(151.5 \text{ kips})(1.67)}{13.5 \text{ in}^2}$$
$$= 18.7 \text{ kips/in}^2 \quad (19 \text{ kips/in}^2)$$

Alternately, the critical compressive stress can be determined using the AISC *Steel Construction Manual*. To prevent localized buckling, AISC *Steel Construction Manual* Sec. B4 limits the width-to-thickness ratio so that the plate element is fully effective. For the double-angle member shown, AISC Table B4.1 gives

$$\frac{b}{t} \leq 0.45\sqrt{\frac{E}{F_y}} = 0.45\sqrt{\frac{29{,}000 \dfrac{\text{kips}}{\text{in}^2}}{36 \dfrac{\text{kips}}{\text{in}^2}}} = 12.77$$

$$\frac{b}{t} = \frac{6 \text{ in}}{0.5 \text{ in}} = 12.0$$

Since 12.0 < 12.77, the section is not classified as a slender element and can be designed in accordance with AISC *Steel Construction Manual* Sec. B and Sec. E3.

AISC *Steel Construction Manual Specification* Sec. E3 gives the requirements for unstiffened compression elements without slender elements such as the angles shown.

The allowable compressive stress depends on Kl/r relative to

$$4.71\sqrt{\frac{E}{F_y}} = 4.71\sqrt{\frac{29{,}000 \dfrac{\text{kips}}{\text{in}^2}}{36 \dfrac{\text{kips}}{\text{in}^2}}} = 134$$

$$\frac{Kl}{r} = 110 \quad [< 134]$$

Therefore, use AISC *Specification* Eq. E3-2 to find the allowable compressive stress.

$$F_{\text{cr}} = \left(0.658^{F_y/F_e}\right)F_y$$

The elastic critical buckling stress, F_e, is determined by AISC *Specification* Eq. E3-4.

$$F_e = \frac{\pi^2 E}{\left(\dfrac{Kl}{r}\right)^2} = \frac{\pi^2 \left(29{,}000 \dfrac{\text{kips}}{\text{in}^2}\right)}{(110)^2} = 23.65 \text{ kips/in}^2$$

[AISC Eq. E3-4]

$$\frac{F_y}{F_e} = \frac{36 \dfrac{\text{kips}}{\text{in}^2}}{23.65 \dfrac{\text{kips}}{\text{in}^2}} = 1.5222$$

$$F_{\text{cr}} = \left(0.658^{F_y/F_e}\right)F_y = \left(0.658^{1.5222}\right)\left(36 \dfrac{\text{kips}}{\text{in}^2}\right)$$
$$= 19.0 \text{ kips/in}^2$$

The answer is (C).

Why Other Options Are Wrong

(A) This incorrect solution finds the stress by dividing the tabulated available strength value by the gross area, neglecting the omega factor.

(B) This incorrect solution solves the problem correctly but identifies the wrong units for the critical stress.

(D) This incorrect solution finds the criticial stress for a member made of ASTM A992 steel instead of ASTM A36 steel.

32. Because the concrete has not gained any of its strength immediately after pouring, the dead loads (including the slab and beam weight) are carried by the steel section alone.

The bending stress in the bottom fibers of the steel beam due to dead load is

$$f = \frac{M}{S}$$

$$f_{\text{bot}} = \frac{M_D}{S_x} = \frac{(80 \text{ ft-kips})\left(12 \dfrac{\text{in}}{\text{ft}}\right)}{68.4 \text{ in}^3}$$
$$= 14 \text{ kips/in}^2$$

The answer is (B).

Why Other Options Are Wrong

(A) This incorrect solution uses the section modulus for the transformed section rather than just for the beam. The transformed section modulus would be used in shored construction.

(C) This incorrect solution uses the total moment instead of just the dead load moment.

(D) This incorrect solution uses ft-lbf for moment instead of in-lbf when calculating the stress. The units do not work out.

33. From ACI 318 Table 9.5(a), determine the minimum thickness, h, for a solid one-way slab with one end continuous.

$$h = \frac{l}{24}$$

l is the clear span length in inches as defined in ACI 318 Sec. 8.7.

$$l = (14 \text{ ft})\left(12 \frac{\text{in}}{\text{ft}}\right) = 168 \text{ in}$$

$$h = \frac{168 \text{ in}}{24} = 7.0 \text{ in}$$

For lightweight concrete, the footnotes to ACI 318 Table 9.5(a) indicate that the tabulated value must be modified.

$$h' = (1.65 - 0.005 w_c)h = \big(1.65 - (0.005)(100)\big)(7.0 \text{ in})$$
$$= 8.1 \text{ in}$$

The answer is (B).

Why Other Options Are Wrong

(A) This incorrect solution neglects to adjust the tabulated value for lightweight concrete. From ACI 318 Table 9.5(a), determine the minimum thickness, h, for a solid one-way slab with one end continuous.

(C) This incorrect solution uses the center-to-center span for l instead of the clear span.

(D) This incorrect solution results from assuming that the slab is simply supported when selecting the thickness equation from ACI 318 Table 9.5(a).

34. ACI 318 Sec. 13.6.5 covers the moments in beams in two-way slab systems.

$$\frac{\alpha_{f1} l_2}{l_1} = \frac{(1.5)(22 \text{ ft})}{24 \text{ ft}} = 1.38$$

Since 1.38 is greater than 1.0, the beam must be proportioned to resist 85% of the column strip moments plus the weight of the beam stem.

$$M_{u,\text{bm}} = 0.85 M_{u,c\,\text{strip}} + M_{u,\text{bm weight}}$$

The midspan moment of a uniformly loaded two-span beam is

$$M_{u,\text{bm weight}} = \frac{w_{u,\text{bm weight}} L^2}{14}$$

$$w_{u,\text{bm weight}} = 1.2 \gamma b h$$
$$= (1.2)\left(150 \frac{\text{lbf}}{\text{ft}^3}\right)(1 \text{ ft})(2 \text{ ft})$$
$$= 360 \text{ lbf/ft}$$

$$M_{u,\text{bm}} = 0.85 M_{u,c\,\text{strip}} + M_{u,\text{bm weight}}$$
$$= (0.85)(90 \text{ ft-kips})$$
$$+ \frac{\left(360 \frac{\text{lbf}}{\text{ft}}\right)\left(\frac{1 \text{ kip}}{1000 \text{ lbf}}\right)(24 \text{ ft})^2}{14}$$
$$= 91 \text{ ft-kips}$$

The answer is (C).

Why Other Options Are Wrong

(A) This incorrect solution does not add the self-weight of the beam to the total moment on the beam.

(B) This incorrect solution does not apply the load factor to the beam weight when calculating the moment on the beam.

(D) This incorrect solution calculates the moment on the beam using the factored slab load and a tributary width equal to 22 ft instead of proportioning the beam moment based on the column strip moment as specified in ACI 318 Sec. 13.6.5.

35. Section 10.3.6 of ACI 318 gives the design axial load strength for compression members. Solve ACI 318 Eq. 10-1 for the gross area of the concrete.

$$\rho_{g_{\max}} = 0.08 \quad [\text{ACI 318 Sec. 10.9.1}]$$

$$\phi = 0.70 \quad \begin{bmatrix} \text{for spiral columns,} \\ \text{ACI 318 Sec. 9.3.2.2} \end{bmatrix}$$

$$P_u = 1.2D + 1.6L$$
$$= (1.2)(300 \text{ kips}) + (1.6)(350 \text{ kips})$$
$$= 920 \text{ kips}$$

$$A_g = \frac{P_u}{0.85\phi\left(0.85f'_c(1-\rho_g)+\rho_g f_y\right)}$$

$$= \frac{(920\text{ kips})\left(1000\ \frac{\text{lbf}}{\text{kip}}\right)}{(0.85)(0.70)\left(\begin{array}{c}(0.85)\left(4000\ \frac{\text{lbf}}{\text{in}^2}\right)(1-0.08)\\ +(0.08)\left(60{,}000\ \frac{\text{lbf}}{\text{in}^2}\right)\end{array}\right)}$$

$$= 195\text{ in}^2 \quad (200\text{ in}^2)$$

The answer is (C).

Why Other Options Are Wrong

(A) This incorrect solution makes a mathematical error in calculating the cross-sectional area by not multiplying the entire denominator by 0.85ϕ.

$$A_g = \frac{P_u}{0.85\phi 0.85f'_c(1-\rho_g)+\rho_g f_y}$$

$$= \frac{(920\text{ kips})\left(1000\ \frac{\text{lbf}}{\text{kip}}\right)}{(0.85)(0.70)(0.85)\left(4000\ \frac{\text{lbf}}{\text{in}^2}\right)(1-0.08)}$$
$$\quad + (0.08)\left(60{,}000\ \frac{\text{lbf}}{\text{in}^2}\right)$$

$$= 138\text{ in}^2 \quad (140\text{ in}^2)$$

(B) This incorrect solution uses $\phi = 0.75$ instead of $\phi = 0.70$.

(D) This incorrect solution uses the adjustment factor for a tied column ($\phi=0.65$) rather than for a spiral column.

36. The design axial load strength of a wall is given in ACI 318 Sec. 14.5.2 as

$$\phi P_n = 0.55\phi f'_c A_g\left(1-\left(\frac{kl_c}{32h}\right)^2\right) \quad \text{[ACI 318 Eq. 14-1]}$$

Calculate the factored axial load, and set it equal to the design axial strength.

$$P_u = w_u l_w = \left(5\ \frac{\text{kips}}{\text{ft}}\right)(16\text{ ft}) = 80\text{ kips}$$

$$P_u = \phi P_n = 0.55\phi f'_c A_g\left(1-\left(\frac{kl_c}{32h}\right)^2\right)$$

$$= 0.55\phi f'_c(hl_w)\left(1-\left(\frac{kl_c}{32h}\right)^2\right)$$

$$\phi = 0.65 \quad \text{[ACI 318 Sec. 14.5.2 and Sec. 9.3.2.2]}$$

$$80\text{ kips} = (0.55)(0.65)\left(4000\ \frac{\text{lbf}}{\text{in}^2}\right)\left(\frac{1\text{ kip}}{1000\text{ lbf}}\right)$$
$$\times (12\text{ in})(16\text{ ft})\left(12\ \frac{\text{in}}{\text{ft}}\right)$$
$$\times \left(1-\left(\frac{(1.0)l_c}{(32)(12\text{ in})}\right)^2\right)$$

$$= (3295\text{ kips})\left(1-\frac{l_c^2}{147{,}456\text{ in}^2}\right)$$

$$0.024282 = 1 - \frac{l_c^2}{147{,}456\text{ in}^2}$$

$$\frac{l_c^2}{147{,}456\text{ in}^2} = 0.975718$$

$$l_c = (379.3\text{ in})\left(\frac{1\text{ ft}}{12\text{ in}}\right)$$
$$= 31.6\text{ ft}$$

Check the minimum thickness requirements found in ACI 318 Sec. 14.5.3. Thickness is based on the shorter of the wall height and wall length. If the wall height is 31.6 ft and the wall length is 16 ft, the minimum thickness is

$$h > \frac{l_w}{25} = \frac{(16\text{ ft})\left(12\ \frac{\text{in}}{\text{ft}}\right)}{25}$$
$$= 7.7\text{ in} \quad \text{[OK]}$$

The answer is (C).

Why Other Options Are Wrong

(A) This incorrect solution does not properly apply the requirements for minimum thickness found in ACI 318 Sec. 14.5.3. The maximum height is incorrectly calculated as the product of 25 times the wall thickness, h. This requirement should be used to check the minimum thickness based upon the lesser of the height or length. In this case, the length, not the height, would be the limiting value.

(B) This incorrect solution calculates the gross area with the length of the wall in feet rather than inches. The units do not work out.

(D) This incorrect solution calculates the uniform axial load, w_u, rather than the total axial load, P_u. The units do not work out.

37. The relative stiffness parameter is the same in each direction for a corner column.

$$\Psi_{\text{E-W}} = \Psi_{\text{N-S}}$$

$$= \frac{\sum\left(\frac{EI}{l_c}\right)_{\text{compr}}}{\sum\left(\frac{EI}{l}\right)_{\text{flex}}} \quad \left[\begin{array}{c}\text{ACI 318}\\ \text{Fig. R10.10.1.1}\end{array}\right]$$

Since E is the same throughout,

$$\Psi = \frac{\sum\left(\frac{I}{l_c}\right)_{\text{compr}}}{\sum\left(\frac{I}{l}\right)_{\text{flex}}}$$

$$I_c = \frac{bh^3}{12} = \frac{(18 \text{ in})^4}{12}$$
$$= 8748 \text{ in}^4$$
$$I_b = 10{,}000 \text{ in}^4$$

Adjust for cracking and creep [ACI 318 Sec. R10.10.4.1].

$$I_{c,e} = 0.70 I_c = (0.70)(8748 \text{ in}^4)$$
$$= 6124 \text{ in}^4$$
$$I_{b,e} = 0.35 I_b = (0.35)(10{,}000 \text{ in}^4)$$
$$= 3500 \text{ in}^4$$

$$\Psi_{\text{top}} = \frac{\sum\left(\frac{I}{l_c}\right)_{\text{compr}}}{\sum\left(\frac{I}{l}\right)_{\text{flex}}} = \frac{\frac{6124 \text{ in}^4}{13 \text{ ft}} + \frac{6124 \text{ in}^4}{18 \text{ ft}}}{\frac{3500 \text{ in}^4}{20 \text{ ft}}}$$
$$= 4.64 \quad (4.6)$$

Because the column is a corner column, there is only one beam in each direction.

The answer is (B).

Why Other Options Are Wrong

(A) This incorrect solution calculates the relative stiffness parameter for an interior column, not a corner column.

(C) This incorrect solution mistakenly uses 13 ft instead of 18 ft for the unbraced column length on the first story.

(D) This incorrect solution does not adjust the moment of inertia of the columns for cracking and creep as recommended in ACI 318 Sec. R10.10.4.1.

38. Section 13.4 of ACI 318 covers openings in slab systems. Without special analysis, openings in a two-way slab system are limited by where they occur.

Determine the width of the column strips in each direction. Column strip width on each side of a column centerline is $0.25 l_1$ or $0.25 l_2$, whichever is less [ACI 318 Sec. 13.2.1].

North/South Direction:

$$l_1 = 20 \text{ ft}$$
$$l_2 = 30 \text{ ft}$$
$$\text{column strip width} = 0.25 l_1 = (0.25)(20 \text{ ft})$$
$$= 5.0 \text{ ft}$$
$$\text{column strip width} = 0.25 l_2 = (0.25)(30 \text{ ft})$$
$$= 7.5 \text{ ft}$$

The lesser value controls.

East/West Direction:

$$l_1 = 30 \text{ ft}$$
$$l_2 = 20 \text{ ft}$$
$$\text{column strip width} = 0.25 l_1 = (0.25)(30 \text{ ft})$$
$$= 7.5 \text{ ft}$$
$$\text{column strip width} = 0.25 l_2 = (0.25)(20 \text{ ft})$$
$$= 5.0 \text{ ft}$$

The lesser value controls.

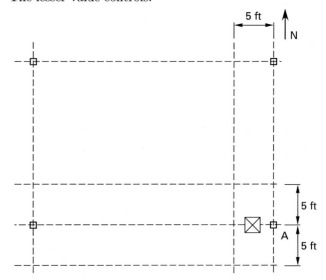

The opening falls within intersecting column strips. ACI 318 Sec. 13.4.2.2 limits openings in intersecting column strips to not more than $1/8$ the width of the column strip in either span.

$$a = b \leq \left(\tfrac{1}{8}\right)(10 \text{ ft}) = 1.25 \text{ ft}$$

The maximum opening is

$$ab = (1.25 \text{ ft})^2 = 1.56 \text{ ft}^2 \quad (1.6 \text{ ft}^2)$$

The answer is (C).

Why Other Options Are Wrong

(A) This incorrect solution assumes that since no special analysis is done, no openings are permitted.

(B) In this incorrect solution, one-half the column strip width is used instead of the full column strip width in calculating the opening size.

(D) This incorrect solution does not use the lesser value in calculating the column strip widths according to ACI 318 Sec. 13.2.1.

39. Since the support condition of the beam is not known, from ACI 318 Sec. 18.4.1, the extreme fiber stress in tension in the concrete is limited to

$$\sigma = 3\sqrt{f'_{ci}} = 3\sqrt{4000 \ \frac{\text{lbf}}{\text{in}^2}}$$
$$= 189.7 \ \text{lbf/in}^2 \quad (190 \ \text{lbf/in}^2)$$

The answer is (B).

Why Other Options Are Wrong

(A) This incorrect solution identifies the initial prestress force in the tendon, P_i, as the tensile stress. Note that the units for initial prestress force are actually kips, not lbf/in^2.

(C) This incorrect solution uses specified strength of concrete rather than the concrete strength at the time of initial prestress when calculating the allowable stress.

(D) This incorrect solution uses the stress limit for simply supported members found in ACI 318 Sec. 18.4.1(c).

40. The nominal shear strength provided by the concrete alone is

$$V_c = 2\sqrt{f'_c}\, b_w d = 2\sqrt{3000 \ \frac{\text{lbf}}{\text{in}^2}}\, (6 \ \text{in})(7.5 \ \text{in})$$
$$= 4929.5 \ \text{lbf}$$
$$\phi V_c = 0.75 V_c = (0.75)(4929.5 \ \text{lbf})$$
$$= 3697 \ \text{lbf}$$

ACI 318 Sec. 11.4.6.1 states that if $V_u < \phi V_c/2$ or $h < 10$ in, then no minimum shear reinforcement is required.

$$\frac{\phi V_c}{2} = \frac{3697 \ \text{lbf}}{2} = 1849 \ \text{lbf}$$
$$V_u = 1800 \ \text{lbf} \quad [< 1849 \ \text{lbf}]$$
$$h = 9 \ \text{in} \quad [< 10 \ \text{in}]$$

Since both conditions are met, no shear reinforcement is required.

The answer is (D).

Why Other Options Are Wrong

(A) This incorrect solution misses the exception, given in ACI 318 Sec. 11.4.6.1, to the minimum shear requirement and mistakenly uses the minimum shear reinforcement requirement found in Sec. 11.4.5.1.

(B) This incorrect solution misses the exception, given in ACI 318 Sec. 11.4.6.1, to the minimum shear requirement and uses h for d when calculating the maximum stirrup spacing.

(C) This incorrect solution misses the exception, given in ACI 318 Sec. 11.4.6.1, to the minimum shear requirement and incorrectly uses the largest stirrup spacing found in ACI 318 Sec. 11.4.5.1 instead of the smallest.

41. Section 11.6 of ACI 318 covers the design of beams with torsion. To determine the maximum stirrup spacing, first calculate the total area of stirrups required, which is the sum of the area required for torsion and the area required for shear.

The transverse reinforcement required for torsion is given by ACI 318 Eq. 11-21.

$$\frac{A_t}{s} = \frac{T_n}{2A_o f_{y,t} \cot\theta} = \frac{\dfrac{T_u}{\phi}}{2A_o f_{y,t} \cot\theta}$$

The area enclosed by the centerline of the torsional reinforcement is

$$A_{oh} = b_t d_t = (31 \ \text{in})(36 \ \text{in}) = 1116 \ \text{in}^2$$
$$A_o = 0.85 A_{oh} = (0.85)(1116 \ \text{in}^2) = 948.6 \ \text{in}^2$$
$$\theta = 45° \quad [\text{ACI 318 Sec. 11.6.3.6}]$$

$$\frac{A_t}{s} = \frac{\dfrac{T_u}{\phi}}{2A_o f_{y,t} \cot\theta} = \frac{\left(\dfrac{400 \ \text{ft-kips}}{0.75}\right)\left(12 \ \dfrac{\text{in}}{\text{ft}}\right)\left(1000 \ \dfrac{\text{lbf}}{\text{kip}}\right)}{(2)(948.6 \ \text{in}^2)\left(40{,}000 \ \dfrac{\text{lbf}}{\text{in}^2}\right)\cot 45°}$$
$$= 0.0843 \ \text{in}^2/\text{in per leg}$$

The reinforcement required for shear is given by ACI 318 Eq. 11-2.

$$V_s = V_n - V_c = \frac{V_u}{\phi} - V_c = \frac{163 \ \text{kips}}{0.75} - 168.2 \ \text{kips}$$
$$= 49.1 \ \text{kips}$$

Using ACI 318 Eq. 11-15,

$$\frac{A_v}{s} = \frac{V_s}{f_{y,t} d} = \frac{(49.1 \ \text{kips})\left(1000 \ \dfrac{\text{lbf}}{\text{kip}}\right)}{\left(40{,}000 \ \dfrac{\text{lbf}}{\text{in}^2}\right)(38 \ \text{in})}$$
$$= 0.0323 \ \text{in}^2/\text{in}$$

Because the area of shear reinforcement is divided between two vertical legs, the total area of stirrups required is

$$A_{\text{req}} = \frac{A_t}{s} + \frac{A_v}{2s} = 0.0843 \, \frac{\text{in}^2}{\text{in}} + \frac{0.0323 \, \frac{\text{in}^2}{\text{in}}}{2}$$
$$= 0.1005 \, \text{in}^2/\text{in}$$

The minimum area of transverse closed stirrups is given by ACI 318 Eq. 11-23.

$$A_v + 2A_t = 0.75\sqrt{f'_c}\left(\frac{b_w s}{f_{y,t}}\right)$$

Rearranging and dividing both sides by $2s$,

$$\frac{1}{2}\left(\frac{2A_t}{s} + \frac{A_v}{s}\right) = \frac{1}{2}\left(\frac{0.75\sqrt{f'_c}\,b_w}{f_{y,t}}\right)$$

$$A_{\min} = \frac{A_t}{s} + \frac{A_v}{2s}$$

$$= \left(\frac{1}{2}\right)\left(\frac{(0.75)\sqrt{4000\,\frac{\text{lbf}}{\text{in}^2}}\,(35\,\text{in})}{40{,}000\,\frac{\text{lbf}}{\text{in}^2}}\right)$$

$$= 0.0208 \, \text{in}^2/\text{in} \quad [< 0.1005 \, \text{in}^2/\text{in}, \text{OK}]$$

Using no. 5 stirrups, A is $0.31 \, \text{in}^2$ per leg.

The spacing required is

$$s = \frac{A}{A_{\text{req}}} = \frac{0.31 \, \text{in}^2}{0.1005 \, \frac{\text{in}^2}{\text{in}}}$$
$$= 3.09 \, \text{in}$$

The maximum spacing permitted by ACI 318 Sec. 11.6.6.1 for torsion is the lesser of $p_h/8$ or 12 in.

p_h is the perimeter of the area enclosed by the centerline of the transverse reinforcement.

$$p_h = 2b_t + 2d_t = (2)(31\,\text{in}) + (2)(36\,\text{in}) = 134\,\text{in}$$
$$\frac{p_h}{8} = \frac{134\,\text{in}}{8} = 16.75\,\text{in}$$

12 in spacing controls for torsion. The maximum spacing permitted by ACI 318 Sec. 11.5.4.1 for shear is the lesser of $d/2$ or 24 in.

$$\frac{d}{2} = \frac{38\,\text{in}}{2} = 19\,\text{in}$$

The 12 in spacing limit controls for both shear and torsion and is greater than the required spacing. Therefore, s is 3.09 in (3.1 in).

The answer is (C).

Why Other Options Are Wrong

(A) This incorrect solution uses the area of a no. 4 stirrup instead of a no. 5 in calculating the required spacing.

(B) This incorrect solution calculates the area considering the maximum torsion and the maximum shear values for the entire span are combined.

(D) This incorrect solution sizes the stirrups for torsion only and does not check the minimum reinforcement requirement found in ACI 318 Sec. 11.6.5.2.

42. ACI 318 Sec. 12.2.2 contains the development length requirements for bars in tension. The development length is measured from the point where the stress in the bars is at maximum. For positive moment, the distance is measured from the center of the span. Since the clear spacing of the bars is greater than twice the diameter of the bars and the clear cover is greater than the diameter of the bars, the development length for a no. 9 bar is given as

$$l_d = \left(\frac{f_y \psi_t \psi_e \lambda}{20\sqrt{f'_c}}\right) d_b$$

$$d_b = 1.128 \, \text{in}$$

ACI 318 Sec. 12.2.4 gives the factors used in this equation.

$$\psi_t = 1.0$$
$$\psi_e = 1.0$$
$$\lambda = 1.0$$

$$l_d = \left(\frac{f_y \psi_t \psi_e \lambda}{20\sqrt{f'_c}}\right) d_b$$

$$= \left(\frac{\left(60{,}000 \, \frac{\text{lbf}}{\text{in}^2}\right)(1.0)(1.0)(1.0)}{20\sqrt{6000 \, \frac{\text{lbf}}{\text{in}^2}}}\right)(1.128\,\text{in})$$

$$= 43.7 \, \text{in}$$

ACI 318 Sec. 12.10.3 states that reinforcement shall extend beyond the point at which it is no longer needed for a distance equal to the greater of the following.

$$d = h - \text{cover} - \frac{d_b}{2} = 30\,\text{in} - 1.5\,\text{in} - \frac{1.128\,\text{in}}{2}$$
$$= 27.9 \, \text{in}$$

$$12 d_b = (12)(1.128\,\text{in}) = 13.5\,\text{in}$$

The point at which the positive moment reinforcement is no longer needed is the point of inflection. Since the point of inflection is 3 ft from the face of the support, the bars must extend to within 36 in − 27.9 in of the face of the support, which equals 8.1 in.

ACI 318 Sec. 12.11.1 states that at least $1/4$ of the positive moment reinforcement must extend into the support at least 6 in. This requirement controls.

The answer is (B).

Why Other Options Are Wrong

(A) This incorrect solution assumes that since there is no positive moment at the support, no reinforcement is required. However, ACI 318 Sec. 12.11.1 indicates otherwise.

(C) This incorrect solution assumes that the point where the positive moment reinforcement is no longer needed is the face of the support. The point of inflection is actually the point where the positive moment goes to zero.

(D) This incorrect solution assumes the development length to be measured from the face of the support. Development length should be measured from the point where the stress in the bars is at maximum. For positive moment, the distance should be measured from the center of the span.

43. Section 2.3 of the NDS contains the adjustment factors and their applicability for design values and references the corresponding sections for each wood member type.

The volume factor, C_V, is discussed in NDS Sec. 5.3 on structural glued laminated timber and in Sec. 8.3 on structural composite lumber. Statement II is true.

The load duration factor, C_D, is discussed in NDS Sec. 2.3.2 and NDS Table 2.3.2. Both NDS Sec. 2.3.2.1 and Footnote 1 to NDS Table 2.3.2 state that "Load duration factors shall not apply to reference modulus of elasticity, E..." Statement IV is true.

The answer is (C).

Why Other Options Are Wrong

(A) Although statement II is true, statement I is false. The temperature factor, C_t, is discussed in NDS Sec. 2.3.3. The temperature factor applies to members subjected to sustained exposure to elevated (over 100°F) temperatures. Because the statement applies to members subjected to extreme cold, not heat, statement I is false.

(B) Although statement II is true, statement III is false. The repetitive member factor, C_r, discussed in Sec. 4.3.9, applies only to dimension lumber bending members 2 in to 4 in thick. Because the joists in this case are 1 in thick, statement III is false.

(D) Although statement IV is true, statement III is false, because the repetitive member factor, C_r, applies only to dimension lumber bending members 2 in to 4 in thick. The joists in this case are 1 in thick.

44. The load on the stringers is

$$w = D + L$$

The weight of the slab is

$$D = t\gamma = (4 \text{ in})\left(\frac{1 \text{ ft}}{12 \text{ in}}\right)\left(150 \; \frac{\text{lbf}}{\text{ft}^3}\right) = 50 \text{ lbf/ft}^2$$

The live load during construction is given as

$$L = 50 \text{ lbf/ft}^2$$

$$w = D + L = 50 \; \frac{\text{lbf}}{\text{ft}^2} + 50 \; \frac{\text{lbf}}{\text{ft}^2} = 100 \text{ lbf/ft}^2$$

The maximum spacing is the lesser of the spacing limited by the bending of the joist, s_j, the spacing limited by the bending of the stringer, s_s, or the spacing limited by the load to the post, s_p.

$$s_j = \sqrt{\frac{4466 \text{ lbf}}{w}} = \sqrt{\frac{4466 \text{ lbf}}{100 \; \frac{\text{lbf}}{\text{ft}^2}}} = 6.68 \text{ ft}$$

$$s_s = \frac{523.4 \; \frac{\text{lbf}}{\text{ft}}}{w} = \frac{523.4 \; \frac{\text{lbf}}{\text{ft}}}{100 \; \frac{\text{lbf}}{\text{ft}^2}} = 5.23 \text{ ft}$$

$$s_p = \frac{512.8 \; \frac{\text{lbf}}{\text{ft}}}{w} = \frac{512.8 \; \frac{\text{lbf}}{\text{ft}}}{100 \; \frac{\text{lbf}}{\text{ft}^2}}$$
$$= 5.13 \text{ ft} \quad (5.1 \text{ ft})$$

The maximum spacing of the stringer is 5.1 ft.

The answer is (A).

Why Other Options Are Wrong

(B) This incorrect solution does not properly convert the slab thickness to feet. A slab thickness of 0.25 ft is used instead of 0.33 ft in calculating the weight of the slab.

(C) This incorrect solution is the largest of the spacings rather than the smallest.

(D) This incorrect solution does not include the live load in the calculation of the load on the stringers, w.

45. The first part of the designation, 20F, indicates the tensile capacity in bending, which is 2000 lbf/in².

The meaning of the combination symbols for glulam timbers can be found in *Standard Specification for Structural Glued Laminated Timber of Softwood Species* (AITC 117) and in textbooks on wood and timber design.

The answer is (D).

Why Other Options Are Wrong

(A) This answer is incorrect. The "V" in the combination symbol indicates that the glulam beam is visually graded. An "E" signifies mechanically graded beams.

(B) This answer is incorrect. The depth of the beam is as designed and is not indicated in the combination symbol.

(C) This answer incorrectly assumes the shear capacity is indicated by V7. The "V" indicates that the member is visually graded. The "7" is part of the combination symbol and is not an indication of strength.

46. NDS Sec. 15.3.2 contains the equations for calculating the column stability factor for built-up columns. The column stability factor is based on the slenderness ratios and must be calculated in each direction. The smaller value is used to determine the allowable compression design value parallel to the grain, F'_c, for the column.

First, determine the dimensional properties of the column.

From NDS Supplement Table 1B, the actual dimensions of 2 × 6 sawn lumber are 1.5 in × 5.5 in.

Using NDS Fig. 15B,

$$d_1 = 5.5 \text{ in}$$
$$d_2 = (3)(1.5 \text{ in}) = 4.5 \text{ in}$$
$$l_1 = (9 \text{ ft})\left(12 \frac{\text{in}}{\text{ft}}\right) = 108 \text{ in}$$
$$l_2 = (4.5 \text{ ft})\left(12 \frac{\text{in}}{\text{ft}}\right) = 54 \text{ in}$$

The column stability factor, C_p, is based on the effective column length, l_e.

$$l_e = K_e l \quad [\text{NDS Sec. 15.3.2.1}]$$

From Table G1 in App. G of the NDS, the buckling length coefficient, K_e, is 1.0 for a column with both ends free to rotate but not free to translate.

$$l_{e1} = K_e l_1 = (1.0)(108 \text{ in}) = 108 \text{ in}$$
$$l_{e2} = K_e l_2 = (1.0)(54 \text{ in}) = 54 \text{ in}$$

Calculate the slenderness ratios. NDS Sec. 15.3.2.3 specifies that the slenderness ratios shall not exceed 50. The slenderness ratios are

$$\frac{l_{e1}}{d_1} = \frac{108 \text{ in}}{5.5 \text{ in}} = 19.6 \quad [<50, \text{OK}]$$
$$\frac{l_{e2}}{d_2} = \frac{54 \text{ in}}{4.5 \text{ in}} = 12.0 \quad [<50, \text{OK}]$$

Calculate the column stability factor in each direction.

$$C_p = K_f \left(\frac{1 + \frac{F_{c,E}}{F_c^*}}{2c} - \sqrt{\left(\frac{1 + \frac{F_{c,E}}{F_c^*}}{2c}\right)^2 - \frac{\frac{F_{c,E}}{F_c^*}}{c}} \right)$$

[NDS Eq. 15.3-1]

$$F_c^* = 1750 \text{ lbf/in}^2 \quad [\text{given}]$$
$$F_{c,E} = \frac{0.822 E'_{min}}{\left(\frac{l_e}{d}\right)^2}$$
$$E'_{min} = 620{,}000 \text{ lbf/in}^2 \quad [\text{given}]$$
$$c = 0.8 \quad [\text{for sawn lumber}]$$

Direction 1:

$$\frac{l_{e1}}{d_1} = 19.6$$

$$K_{f1} = 1.0 \quad \left[\begin{array}{c}\text{for a built-up column where } l_{e1}/d_1 \\ \text{is used to calculate } F_{c,E}\end{array}\right]$$

$$F_{c,E_1} = \frac{0.822 E'_{min}}{\left(\frac{l_{e1}}{d_1}\right)^2} = \frac{(0.822)\left(620{,}000 \frac{\text{lbf}}{\text{in}^2}\right)}{(19.6)^2}$$
$$= 1327 \text{ lbf/in}^2$$

$$\frac{F_{c,E_1}}{F_c^*} = \frac{1327 \frac{\text{lbf}}{\text{in}^2}}{1750 \frac{\text{lbf}}{\text{in}^2}} = 0.758$$

$$C_{p1} = K_f \left(\frac{1 + \frac{F_{c,E}}{F_c^*}}{2c} - \sqrt{\left(\frac{1 + \frac{F_{c,E}}{F_c^*}}{2c}\right)^2 - \frac{\frac{F_{c,E}}{F_c^*}}{c}} \right)$$

$$= (1.0)\left(\frac{1 + 0.758}{(2)(0.8)} - \sqrt{\left(\frac{1 + 0.758}{(2)(0.8)}\right)^2 - \frac{0.758}{0.8}}\right)$$

$$= 0.59$$

Direction 2:

$$\frac{l_{e2}}{d_2} = 12.0$$

$$K_{f2} = 0.6 \quad \left[\begin{array}{c}\text{for a built-up column where } l_{e2}/d_2 \text{ is used to} \\ \text{calculate } F_{c,E} \text{ and the column is nailed}\end{array}\right]$$

$$F_{c,E_2} = \frac{0.822 E'_{\min}}{\left(\dfrac{l_{e2}}{d_2}\right)^2} = \frac{(0.822)\left(620{,}000\ \dfrac{\text{lbf}}{\text{in}^2}\right)}{(12.0)^2}$$

$$= 3539\ \text{lbf/in}^2$$

$$\frac{F_{c,E_2}}{F_c^*} = \frac{3539\ \dfrac{\text{lbf}}{\text{in}^2}}{1750\ \dfrac{\text{lbf}}{\text{in}^2}} = 2.0$$

$$C_{p2} = K_f\left(\frac{1+\dfrac{F_{c,E}}{F_c^*}}{2c} - \sqrt{\left(\dfrac{1+\dfrac{F_{c,E}}{F_c^*}}{2c}\right)^2 - \dfrac{\dfrac{F_{c,E}}{F_c^*}}{c}}\right)$$

$$= (0.6)\left(\frac{1+2.0}{(2)(0.8)} - \sqrt{\left(\frac{1+2.0}{(2)(0.8)}\right)^2 - \frac{2.0}{0.8}}\right)$$

$$= 0.52$$

The critical column stability factor is the smaller of C_{p1} and C_{p2}, which is 0.52.

The answer is (B).

Why Other Options Are Wrong

(A) This incorrect solution reverses the dimensions of the column, d_1 and d_2, in calculating the slenderness ratios.

(C) This incorrect solution uses a buckling length coefficient, K_e, of 0.65 from NDS Table G1 in App. G for a column with rotation and translation fixed at both ends.

(D) This incorrect solution uses the K_{f2} for bolted columns instead of nailed columns. It could also be arrived at by calculating C_p correctly but selecting the larger value as the critical one.

47. Section 11.2.3 of the NDS gives the withdrawal values for a single nail. From NDS Table 11.3.2A, the specific gravity of southern pine is 0.55.

NDS Table 11P gives the diameter of a 6d box nail as 0.099 in.

From NDS Table 11.2C, the tabulated withdrawal design value for a specific gravity of 0.55 and a nail diameter of 0.099 in is 31 lbf per inch of penetration.

The design withdrawal value is

$C_D = 1.6$ [for wind load combinations]

$C_M = 1.0$ [for moisture content $\leq 19\%$]

$C_t = 1.0$ [according to NDS Table 10.3.4]

$C_{\text{tn}} = 1.0$ [for non-toe-nailed connections]

$$W' = W C_D C_M C_t C_{\text{tn}}$$
$$= (31\ \text{lbf})(1.6)(1.0)(1.0)(1.0)$$
$$= 49.6\ \text{lbf}$$

Calculate the maximum area per nail.

$$A = \frac{W'}{w} = \frac{49.6\ \text{lbf}}{36\ \dfrac{\text{lbf}}{\text{ft}^2}} = 1.38\ \text{ft}^2$$

If the rafters are spaced at 16 in on center,

$$s_r = (16\ \text{in})\left(\frac{1\ \text{ft}}{12\ \text{in}}\right) = 1.33\ \text{ft}$$

The maximum nail spacing is

$$s_n = \frac{A}{s_r} = \left(\frac{1.38\ \text{ft}^2}{1.33\ \text{ft}}\right)\left(12\ \frac{\text{in}}{\text{ft}}\right)$$

$$= 12.4\ \text{in} \quad (12\ \text{in})$$

The answer is (C).

Why Other Options Are Wrong

(A) This incorrect solution neglects the load duration factor in calculating the withdrawal capacity.

(B) This incorrect solution uses a rafter spacing of 1.5 ft instead of 1.33 ft.

(D) This incorrect solution bases the tabulated withdrawal value on the diameter of a 6d common nail (0.113 in) instead of a 6d box nail.

48. The beam-to-post connection shown imparts a lateral load on the wood screws. Because two wood members are joined in this connection, the connection is subjected to single shear. Appendix I of the NDS illustrates connections in single and double shear. Use NDS Chap. 10 and Chap. 11 and NDS Table 11L to determine the allowable lateral load on the connection.

The allowable lateral design value for a single connector is given in NDS Table 10.3.1 as

$$Z' = Z C_D C_M C_t C_g C_\Delta C_{\text{eg}} C_{di} C_{\text{tn}}$$

In this case, C_D is 1.0 for dead load plus live load, since the load duration factor for the shortest duration load applies [NDS Sec. 2.3.2 and App. B]. C_M is 0.7, since the moisture content of this deck built in a humid climate will exceed 19% [NDS Table 10.3.3]. Values for C_g and C_Δ depend on the diameter of the screw, D. Since $D < 0.25$ in for a 12-gage screw ($D = 0.216$ in), C_g and C_Δ equal 1.0 [NDS Sec. 10.3.6.1 and Sec. 11.5.1]. C_{eg}, C_{di}, and C_{tn} do not apply.

Therefore,

$$Z' = ZC_DC_MC_gC_\Delta = Z(1.0)(0.7)(1.0)(1.0)$$

The cut-thread wood screw design values are given in NDS Table 11L. First, determine the side-member thickness. The side member is the 2 × 8 beam. Using Table 1B of the NDS Supplement, the side-member thickness is 1.5 in. Next, find the column for southern pine and the row for a 12-gage wood screw. The tabulated design value is

$$Z = 160 \text{ lbf}$$
$$Z' = ZC_DC_MC_gC_\Delta = (160 \text{ lbf})(1.0)(0.7)(1.0)(1.0)$$
$$= 112 \text{ lbf} \quad [\text{per screw}]$$

The capacity of the connection is

$$Z'_{\text{total}} = (5 \text{ screws})\left(112 \frac{\text{lbf}}{\text{screw}}\right) = 560 \text{ lbf}$$

The answer is (C).

Why Other Options Are Wrong

(A) This incorrect solution misreads NDS Table 11L and uses the value for a 10-gage screw ($Z = 127$ lbf) instead of the 12-gage screw value.

(B) This incorrect solution uses the wrong load duration factor. The load duration factor for the shortest-duration load applies. This solution uses the smallest load duration factor, $C_D = 0.9$.

(D) This incorrect solution fails to apply the adjustment factors to the tabulated design value to determine the allowable design value.

49. The span length of the lintel is

$$L = \text{clear span} + \frac{1}{2}\sum \text{bearing length at each end}$$
$$= 6.0 \text{ ft} + \left(\frac{1}{2}\right)(8 \text{ in} + 8 \text{ in})\left(\frac{1 \text{ ft}}{12 \text{ in}}\right)$$
$$= 6.67 \text{ ft}$$

Calculate the loads on the lintel.

The total uniform load is the sum of the dead and live loads plus the lintel self-weight.

$$w_{\text{uniform}} = w_D + w_L + w_{\text{self-wt}}$$
$$= 450 \frac{\text{lbf}}{\text{ft}} + 550 \frac{\text{lbf}}{\text{ft}} + 120 \frac{\text{lbf}}{\text{ft}}$$
$$= 1120 \text{ lbf/ft}$$

The total triangular load is

$$W = \tfrac{1}{2}Lh = \left(\tfrac{1}{2}\right)(6.67 \text{ ft})\left(300 \frac{\text{lbf}}{\text{ft}}\right)$$
$$= 1000 \text{ lbf}$$

Calculate the design moment.

$$M_{\text{total}} = M_{\text{uniform}} + M_{\text{triangular}} = \frac{wL^2}{8} + \frac{WL}{6}$$
$$= \left(\frac{\left(1120 \frac{\text{lbf}}{\text{ft}}\right)(6.67 \text{ ft})^2}{8}\right)\left(12 \frac{\text{in}}{\text{ft}}\right)$$
$$+ \left(\frac{(1000 \text{ lbf})(6.67 \text{ ft})}{6}\right)\left(12 \frac{\text{in}}{\text{ft}}\right)$$
$$= 88{,}081 \text{ in-lbf}$$

Determine the effective depth to reinforcement, d. The two bars side-by-side at the bottom of the beam are shown.

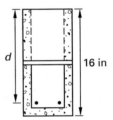

Allowing for the thickness of the unit (1.25 in) and one-half the bar diameter (0.25 in) plus grout cover (0.5 in), estimate d.

$$d = 16 \text{ in} - (1.25 \text{ in} + 0.25 \text{ in} + 0.5 \text{ in}) = 14 \text{ in}$$

Determine the stress in the steel using the equations found in a reference handbook on masonry walls.

$$f_s = \frac{M}{A_sjd}$$
$$\rho = \frac{A_s}{bd}$$
$$k = \sqrt{2\rho n + (\rho n)^2} - \rho n$$
$$j = 1 - \frac{k}{3}$$
$$n = \frac{E_s}{E_m}$$

Building Code Requirements for Masonry Structures (MSJC) Sec. 1.8 specifies the moduli of elasticity of steel and masonry, respectively, as

$$E_s = 29 \times 10^6 \text{ lbf/in}^2$$

$$E_m = 900 f'_m \quad \text{[for concrete masonry]}$$
$$= (900)\left(3000 \; \frac{\text{lbf}}{\text{in}^2}\right)$$
$$= 2.7 \times 10^6 \; \text{lbf/in}^2$$
$$n = \frac{E_s}{E_m} = \frac{29 \times 10^6 \; \frac{\text{lbf}}{\text{in}^2}}{2.7 \times 10^6 \; \frac{\text{lbf}}{\text{in}^2}} = 10.7$$

The allowable tensile stress for grade 60 reinforcement is given in *MSJC* Sec. 2.3.

$$F_s = 24{,}000 \; \text{lbf/in}^2$$

Given two no. 4 bars, App. E of ACI 318 lists the area of a no. 4 reinforcing bar as 0.20 in².

$$A_{s_{\text{prov}}} = (2)(0.20 \; \text{in}^2) = 0.40 \; \text{in}^2$$
$$\rho = \frac{A_{s_{\text{prov}}}}{bd} = \frac{0.40 \; \text{in}^2}{(7.5 \; \text{in})(14 \; \text{in})}$$
$$= 0.00381$$

$$k = \sqrt{2\rho n + (\rho n)^2} - \rho n$$
$$= \sqrt{(2)(0.00381)(10.7) + \big((0.00381)(10.7)\big)^2}$$
$$\quad - (0.00381)(10.7)$$
$$= 0.248$$
$$j = 1 - \frac{k}{3} = 1 - \frac{0.248}{3} = 0.917$$

The stress in the steel is

$$f_s = \frac{M}{A_s j d} = \frac{88{,}080 \; \text{in-lbf}}{(0.40 \; \text{in}^2)(0.917)(14 \; \text{in})}$$
$$= 17{,}152 \; \text{lbf/in}^2 \quad (17{,}000 \; \text{lbf/in}^2)$$

The allowable tensile stress in the steel is

$$F_s = 24{,}000 \; \text{lbf/in}^2 \quad [> f_s]$$

Therefore, the tensile stress in the steel is 17,000 lbf/in².

The answer is (D).

Why Other Options Are Wrong

(A) This incorrect solution uses the clear span for the span length. The span length should include one-half of the bearing length on each side of the opening.

(B) This incorrect solution uses the depth of the lintel (16 in) for d instead of the effective depth to the reinforcement.

(C) This incorrect solution disregards the self-weight of the lintel in calculating the loads.

50. The necessary equations for calculating the stresses in masonry can be found in a masonry handbook and in *Building Code Requirements for Masonry Structures* (*MSJC*).

The allowable moment based on flexural compressive stress in the masonry is

$$M_m = \tfrac{1}{2} F_b b d^2 j k$$
$$F_b = \tfrac{1}{3} f'_m \quad [\textit{MSJC} \; \text{Sec. 2.3.3}]$$

The allowable moment based on flexural tensile stress in the steel is

$$M_s = A_s F_s j d$$

For grade 60 reinforcement,

$$F_s = 24{,}000 \; \text{lbf/in}^2 \quad [\textit{MSJC} \; \text{Sec. 2.3.2}]$$

$$\rho = \frac{A_s}{bd}$$
$$k = \sqrt{2\rho n + (\rho n)^2} - \rho n$$
$$j = 1 - \frac{k}{3}$$
$$n = \frac{E_s}{E_m} = 21.5 \quad \text{[given]}$$

Assume the vertical reinforcement is centered in the middle of the wall. The effective depth to reinforcement is

$$d = \frac{t}{2} = \frac{5.63 \; \text{in}}{2} = 2.8 \; \text{in}$$

Determine the required area of reinforcement.

$$A_{s,\text{req}} = \frac{M_{\max}}{F_s j d}$$

The maximum moment on the wall is

$$M_{\max} = \left(341 \; \frac{\text{ft-lbf}}{\text{ft}}\right)\left(12 \; \frac{\text{in}}{\text{ft}}\right) = 4092 \; \text{in-lbf/ft}$$

Estimate that j is equal to 0.9.

$$A_{s,\text{req}} = \frac{M_{\max}}{F_s j d} = \frac{4092 \; \frac{\text{in-lbf}}{\text{ft}}}{\left(24{,}000 \; \frac{\text{lbf}}{\text{in}^2}\right)(0.9)(2.8 \; \text{in})}$$
$$= 0.0677 \; \text{in}^2/\text{ft}$$

The problem statement gives the bar spacing as 32 in. A no. 4 bar has an area of 0.20 in². Try using no. 4 bars at 32 in.

$$A_{s,\text{prov}} = \left(\frac{0.20 \; \text{in}^2}{32 \; \text{in}}\right)\left(12 \; \frac{\text{in}}{\text{ft}}\right) = 0.075 \; \text{in}^2/\text{ft}$$

MSJC Sec. 1.9.6.1 limits the effective compressive width per bar to the smallest of

(a) center-to-center bar spacing: 32 in

(b) six times the nominal wall thickness: (6)(6 in) = 36 in

(c) 72 in

Therefore, the effective compressive width per bar is the center-to-center bar spacing, or 32 in. Calculate the stresses based on this effective width.

To check the masonry stresses, first calculate j, k, and ρ. For a 32 in length of wall,

$$A_s = 0.20 \text{ in}^2$$

$$\rho = \frac{A_s}{bd} = \frac{0.20 \text{ in}^2}{(32 \text{ in})(2.8 \text{ in})}$$

$$= 0.00223$$

$$n = \frac{E_s}{E_m} = 21.5 \quad \text{[given]}$$

$$k = \sqrt{2\rho n + (\rho n)^2} - \rho n$$

$$= \sqrt{(2)(0.00223)(21.5) + \big((0.00223)(21.5)\big)^2}$$

$$\quad - (0.00223)(21.5)$$

$$= 0.265$$

$$j = 1 - \frac{k}{3} = 1 - \frac{0.265}{3} = 0.912$$

When the neutral axis falls within the face shell, kd is less than the face-shell thickness, and the analysis is the same as for a fully grouted masonry wall. See the *Structural Engineering Reference Manual* for further discussion.

$$kd = (0.265)(2.8 \text{ in}) = 0.74 \text{ in} \quad [< 1.0 \text{ in face shell, OK}]$$

Calculate the allowable moment based on flexural compressive stress in the masonry for a 32 in length of wall.

$$F_b = \tfrac{1}{3}f'_m = \left(\tfrac{1}{3}\right)\left(1500 \, \tfrac{\text{lbf}}{\text{in}^2}\right)$$

$$= 500 \text{ lbf/in}^2$$

$$M_m = \tfrac{1}{2}F_b bd^2 jk$$

$$= \left(\tfrac{1}{2}\right)\left(500 \, \tfrac{\text{lbf}}{\text{in}^2}\right)(32 \text{ in})(2.8 \text{ in})^2(0.912)(0.265)$$

$$= 15{,}158 \text{ in-lbf} \quad (15{,}000 \text{ in-lbf})$$

$$M_{\max} = \left(4092 \, \tfrac{\text{in-lbf}}{\text{ft}}\right)(32 \text{ in})\left(\tfrac{1 \text{ ft}}{12 \text{ in}}\right)$$

$$= 10{,}912 \text{ in-lbf}$$

$$M_m > M_{\max} \quad \text{[OK]}$$

Confirm the allowable moment based on the flexural tensile stress in the steel.

$$M_s = A_s F_s j d$$

$$= \left(0.075 \, \tfrac{\text{in}^2}{\text{ft}}\right)\left(24{,}000 \, \tfrac{\text{lbf}}{\text{in}^2}\right)(0.912)(2.8 \text{ in})$$

$$= 4596 \text{ in-lbf/ft} \quad [> M_{\max}, \text{ OK}]$$

The allowable moment per foot based on the flexural compressive stress in the masonry is

$$M_m = (15{,}158 \text{ in-lbf})\left(\frac{12 \, \tfrac{\text{in}}{\text{ft}}}{32 \text{ in}}\right)$$

$$= 5684 \text{ in-lbf/ft} \quad (5700 \text{ in-lbf/ft})$$

The answer is (C).

Why Other Options Are Wrong

(A) This incorrect solution gives the maximum moment on the wall.

(B) This incorrect solution gives the answer for the allowable moment based on the tensile stress in the steel (4600 in-lbf/ft) instead of the allowable compressive flexural moment (5700 in-lbf/ft).

(D) This incorrect solution uses the $\tfrac{1}{3}$ allowable stress increase for wind when determining the amount of reinforcement required and checking the stresses in the masonry. The $\tfrac{1}{3}$ stress increase is no longer permitted by code.

51. Since the compressive strength of masonry is not specified and material properties are not given, consider empirical design. Chap. 5 of *Building Code Requirements for Masonry Structures* (*MSJC*) contains the empirical requirements for thickness. Two requirements must be checked: one for lateral support and one for minimum thickness. The greater value applies.

MSJC Table 5.5.1 gives wall lateral support requirements. Fully grouted bearing walls have a maximum length-to-thickness ratio, l/t, or height-to-thickness ratio, h/t, of 20. The limiting length or height is the distance between lateral supports. If the wall spans horizontally, the limiting ratio is the length-to-thickness, l/t. If the wall spans vertically, the limiting ratio is the height-to-thickness, h/t.

Since the wall spans vertically, $h/t \leq 20$.

Solving for the thickness,

$$t \geq \frac{h}{20} = \frac{(10 \text{ ft})\left(12 \, \tfrac{\text{in}}{\text{ft}}\right)}{20}$$

$$= 6.0 \text{ in}$$

The minimum thickness requirements of *MSJC* Sec. 5.6.2 also apply. This section specifies the minimum wall thickness based on the number of stories. Since the problem states that the wall spans 10 ft from the foundation to the roof, assume the wall is one story. One-story walls must have a minimum thickness of 6 in.

The answer is (A).

Why Other Options Are Wrong

(B) This incorrect solution does not recognize the wall described as a single story and uses the minimum thickness for walls more than one story (8 in).

(C) This incorrect solution uses the wall's length-to-thickness ratio as the limiting ratio. Although lateral support is provided by the intersecting walls, the problem states that the wall spans vertically. The span supports at the foundation and roof provide lateral support as well. The limiting ratio should be based on the direction of span, h/t.

(D) This incorrect solution does not recognize the wall described as being an empirically designed wall. Walls designed using allowable stress design (ASD) do not have limits on their thicknesses.

The fact that the compressive strength of masonry, f'_m, is not specified identifies this wall as empirically designed. Walls designed using allowable stress design must specify f'_m.

52. Chapter 6 of the NDS contains the specifications for round timber piles. Using ASD, the allowable compression design value is

$$F'_c = F_c C_D C_t C_u C_p C_{cs}$$

The reference compression design values, F_c, apply to pile clusters. The single pile factor, C_{sp}, only applies to piles carrying individual loads and so was omitted.

From NDS Table 2.3.3, for in-service temperatures less than 100°F,

$$C_t = 1.0$$

From NDS Table 6.3.5, for kiln-dried piles made of red oak,

$$C_u = 1.11$$
$$C_p = 0.62 \quad \text{[given]}$$

From NDS Sec. 6.3.9, for round timber piles made of red oak,

$$C_{cs} = 1.0$$
$$F_c = 1100 \text{ lbf/in}^2 \quad \text{[NDS Table 6A]}$$

The load duration factor depends upon the type of load. The load duration factor for the shortest duration load in a combination of loads shall apply for that load combination [NDS Sec. 2.3.2.2]. There are two load cases to consider: dead load and dead load plus live load.

Dead Load Only:

$$C_D = 0.9$$
$$F'_c = F_c C_D C_t C_u C_p C_{cs}$$
$$= \left(1100 \frac{\text{lbf}}{\text{in}^2}\right)(0.9)(1.0)(1.11)(0.62)(1.0)$$
$$= 681 \text{ lbf/in}^2$$

Dead Load plus Live Load:

$$C_D = 1.0$$
$$F'_c = F_c C_D C_t C_u C_p C_{cs}$$
$$= \left(1100 \frac{\text{lbf}}{\text{in}^2}\right)(1.0)(1.0)(1.11)(0.62)(1.0)$$
$$= 757 \text{ lbf/in}^2$$

Determine the critical load case.

Dead Load Only:

$$f_{c,D} = \frac{P_D}{A} = \frac{\left(\dfrac{300 \text{ kips}}{3 \text{ piles}}\right)\left(1000 \dfrac{\text{lbf}}{\text{kip}}\right)}{230 \text{ in}^2}$$
$$= 435 \text{ lbf/in}^2 \text{ per pile} \quad [< F'_c, \text{ OK}]$$

Dead Load plus Live Load:

$$f_{c,D+L} = \frac{P_{D+L}}{A}$$
$$= \frac{\left(\dfrac{300 \text{ kips} + 400 \text{ kips}}{3 \text{ piles}}\right)\left(1000 \dfrac{\text{lbf}}{\text{kip}}\right)}{230 \text{ in}^2}$$
$$= 1014 \text{ lbf/in}^2 \text{ per pile} \quad [> F'_c, \text{ NG}]$$

The critical load case is dead load plus live load. The total adjustment factor for this load case is 0.688.

The answer is (D).

Why Other Options Are Wrong

(A) This incorrect solution includes the single pile factor [NDS Sec. 6.3.12]. Although the stress is calculated for a single pile, the pile is still one in a group, and this factor does not apply.

(B) This incorrect solution uses the *smallest* load duration factor (0.9) for the dead plus live load combination instead of using the load duration factor for the *shortest* load (1.0).

(C) This incorrect solution does not include the untreated factor, C_u.

53. The sliding force is resisted by friction and adhesion between the soil and the base. Passive restraint from the soil is only considered if it will always be there. In most cases, it is neglected. The sliding force is the sum of the pressure due to the earth and the pressure due to the slab surcharge. From the *Structural Engineering Reference Manual* or another reference book,

$$R_{a,h} = \tfrac{1}{2} k_a \gamma H^2 + k_a q H$$

The lateral earth pressure due to the soil backfill is

$$R_{a,h,\text{soil}} = \tfrac{1}{2} k_a \gamma H^2 = \left(\tfrac{1}{2}\right)(0.361)\left(0.110 \ \tfrac{\text{kip}}{\text{ft}^3}\right)(20 \text{ ft})^2$$
$$= 7.94 \text{ kips/ft}$$

The lateral earth pressure due to the surcharge from the concrete slab is

$$R_{a,h,\text{surcharge}} = k_a q H$$
$$= (0.361)\left(\begin{array}{c}(10 \text{ in})\left(\tfrac{1 \text{ ft}}{12 \text{ in}}\right)\\ \times \left(0.150 \ \tfrac{\text{kip}}{\text{ft}^3}\right)\end{array}\right)(20 \text{ ft})$$
$$= 0.90 \text{ kip/ft}$$

$$R_{a,h} = \tfrac{1}{2} k_a \gamma H^2 + k_a q H = 7.94 \ \tfrac{\text{kips}}{\text{ft}} + 0.90 \ \tfrac{\text{kip}}{\text{ft}}$$
$$= 8.84 \text{ kips/ft}$$

Sliding is resisted by friction from the weight of the soil, retaining wall, and surcharge. From a reference handbook such as the *Structural Engineering Reference Manual*,

$$R_{\text{SL}} = \left(\sum W_i + R_{a,v}\right) \tan \delta + c_A B$$
$$W_i = \gamma_i A_i$$

Since the backfill has no slope, there is no vertical component of the active earth pressure.

$$R_{a,v} = 0$$

According to IBC Table 1804.2, without other information, $\tan \delta$ should be taken as 0.35 for sandy soil without silt.

$$c_A = 0 \quad [\text{for granular soil}]$$
$$R_{\text{SL}} = \left(\sum W_i\right) \tan \delta$$
$$= \left(\sum W_i\right)(0.35)$$

Determine the weight of the soil, wall, and surcharge.

Soil:

$$A_{\text{soil}} = LH = \left(15 \text{ ft} - (20 \text{ in})\left(\tfrac{1 \text{ ft}}{12 \text{ in}}\right)\right)(20 \text{ ft} - 2 \text{ ft})$$
$$= 240 \text{ ft}^2$$
$$W_{\text{soil}} = \gamma_{\text{soil}} A_{\text{soil}} = \left(0.110 \ \tfrac{\text{kip}}{\text{ft}^3}\right)(240 \text{ ft}^2)$$
$$= 26.4 \text{ kips/ft}$$

Retaining Wall, Vertical Section:

$$A_{w,v} = LH = (20 \text{ in})\left(\tfrac{1 \text{ ft}}{12 \text{ in}}\right)(20 \text{ ft}) = 33.3 \text{ ft}^2$$
$$W_{w,v} = \gamma_w A_w = \left(0.150 \ \tfrac{\text{kip}}{\text{ft}^3}\right)(33.3 \text{ ft}^2) = 5.00 \text{ kips/ft}$$

Retaining Wall, Horizontal Section:

$$A_{w,h} = LH = (13.33 \text{ ft})(2 \text{ ft})$$
$$= 26.7 \text{ ft}^2$$
$$W_{w,h} = \gamma_w A_w = \left(0.150 \ \tfrac{\text{kip}}{\text{ft}^3}\right)(26.7 \text{ ft}^2)$$
$$= 4.01 \text{ kips/ft}$$

Concrete Slab Surcharge:

$$A_{\text{slab}} = LH = (8 \text{ ft})(10 \text{ in})\left(\tfrac{1 \text{ ft}}{12 \text{ in}}\right)$$
$$= 6.67 \text{ ft}^2$$
$$W_{\text{slab}} = \gamma_{\text{slab}} A_{\text{slab}} = \left(0.150 \ \tfrac{\text{kip}}{\text{ft}^3}\right)(6.67 \text{ ft}^2)$$
$$= 1.00 \text{ kip/ft}$$

$$\sum W_i = W_{\text{soil}} + W_{w,v} + W_{w,h} + W_{\text{slab}}$$
$$= 26.4 \ \tfrac{\text{kips}}{\text{ft}} + 5.00 \ \tfrac{\text{kips}}{\text{ft}}$$
$$+ 4.01 \ \tfrac{\text{kips}}{\text{ft}} + 1.00 \ \tfrac{\text{kip}}{\text{ft}}$$
$$= 36.4 \text{ kips/ft}$$

The resistance against sliding is

$$R_{\text{SL}} = \left(\sum W_i\right) \tan \delta = \left(36.4 \ \tfrac{\text{kips}}{\text{ft}}\right)(0.35)$$
$$= 12.7 \text{ kips/ft}$$

The factor of safety against sliding is

$$F_{\text{SL}} = \frac{R_{\text{SL}}}{R_{a,h}} = \frac{12.7 \ \tfrac{\text{kips}}{\text{ft}}}{8.84 \ \tfrac{\text{kips}}{\text{ft}}} = 1.44$$

The answer is (B).

Why Other Options Are Wrong

(A) In this incorrect solution, the weight of the slab is not included when calculating the resisting force.

(C) This incorrect solution neglects the effect of the surcharge entirely.

(D) This incorrect solution includes the restraint provided by the passive force. Unless explicitly stated that it will always be there, passive restraint should be ignored.

54. The minimum thickness of a footing is determined by the greater of the minimum depth required for shear and the minimum depth required by Sec. 15.7 of ACI 318.

The required depth for shear is based on the ultimate soil pressure. The ultimate soil pressure is

$$q_u = \frac{1.2P_D + 1.6P_L}{A}$$
$$= \frac{(1.2)(50 \text{ kips}) + (1.6)(75 \text{ kips})}{(6.5 \text{ ft})(6.5 \text{ ft})}$$
$$= 4.26 \text{ kips/ft}^2$$

From ACI 318 Sec. 9.3.2, determine that the strength reduction factor, ϕ, for shear is 0.75.

From Chap. 11 of ACI 318, the allowable shear stress in the concrete footing, assuming two-way action, is

$$v_c = 4\phi\sqrt{f'_c}$$
$$= (4)(0.75)\sqrt{3000 \, \frac{\text{lbf}}{\text{in}^2}} \left(\frac{1 \text{ kip}}{1000 \text{ lbf}}\right) \left(12 \, \frac{\text{in}}{\text{ft}}\right)^2$$
$$= 23.7 \text{ kips/ft}^2$$

Sum the shear forces acting on the footing to determine the footing depth. For a concrete footing for a square concrete column, the equation becomes

$$d^2\left(v_c + \frac{q_u}{4}\right) + d\left(v_c + \frac{q_u}{2}\right)w = (B^2 - w^2)\left(\frac{q_u}{4}\right)$$

$$d^2\left(23.7 \, \frac{\text{kips}}{\text{ft}^2} + \frac{4.26 \, \frac{\text{kips}}{\text{ft}^2}}{4}\right) + d\left(\begin{array}{c} 23.7 \, \frac{\text{kips}}{\text{ft}^2} \\ + \frac{4.26 \, \frac{\text{kips}}{\text{ft}^2}}{2} \end{array}\right)w$$
$$= ((6.5 \text{ ft})^2 - (1.0 \text{ ft})^2) \left(\frac{4.26 \, \frac{\text{kips}}{\text{ft}^2}}{4}\right)$$

$$d^2\left(24.77 \, \frac{\text{kips}}{\text{ft}^2}\right) + d\left(25.83 \, \frac{\text{kips}}{\text{ft}^2}\right)w = 43.93 \text{ kips}$$

Using the quadratic equation and solving for d gives

$$d = (0.91 \text{ ft})\left(12 \, \frac{\text{in}}{\text{ft}}\right) = 10.9 \text{ in}$$

The minimum thickness of the footing is

$$h = d + \tfrac{1}{2}(d_{b,x} + d_{b,y}) + \text{cover}$$

A no. 5 bar has a diameter of 0.625 in [ACI 318 App. E]. The minimum cover (below the bars) for concrete against the earth is 3 in [ACI 318 Sec. 7.7.1].

$$h = d + \tfrac{1}{2}(d_{b,x} + d_{b,y}) + \text{cover}$$
$$= 10.9 \text{ in} + \left(\tfrac{1}{2}\right)(0.625 \text{ in} + 0.625 \text{ in}) + 3 \text{ in}$$
$$= 14.5 \text{ in} \quad (15 \text{ in}) \quad \begin{bmatrix} \text{Round up to 15 in} \\ \text{because } h \text{ must be} \\ \geq 14.5 \text{ in.} \end{bmatrix}$$

Check that this depth exceeds the minimum required by the code. ACI 318 Sec. 15.7 specifies that the depth of a footing above the bottom of reinforcement must be greater than 6 in.

$$d = 10.9 \text{ in} \quad [> 6 \text{ in, OK}]$$

The answer is (D).

Why Other Options Are Wrong

(A) This incorrect solution does not convert the allowable shear stress to kips/ft^2 in the calculation of depth to reinforcement.

(B) This incorrect solution neglects to add cover to determine the overall footing depth.

(C) This incorrect solution uses unfactored loads. Concrete should be designed using ultimate strength equations and factored loads.

55. The size of the footing is based on service (unfactored) loads and soil pressures, because footing design safety is provided by the safety factor in the allowable soil bearing pressure.

$$P = P_D + P_L$$
$$M = M_D + M_L$$

For column 1,

$$P_1 = 30 \text{ kips} + 60 \text{ kips} = 90 \text{ kips}$$
$$M_1 = 30 \text{ ft-kips} + 40 \text{ ft-kips} = 70 \text{ ft-kips}$$

For column 2,

$$P_2 = 60 \text{ kips} + 60 \text{ kips} = 120 \text{ kips}$$
$$M_2 = 30 \text{ ft-kips} + 40 \text{ ft-kips} = 70 \text{ ft-kips}$$

For the soil pressure under the footing to be uniform, the resultant load, R, must be located at the centroid of the base area.

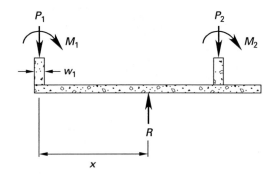

Summing moments about column 1, x is the distance to the centroid of the footing.

$$Rx = P_2(18 \text{ ft}) + M_1 + M_2$$
$$(P_1 + P_2)x = P_2(18 \text{ ft}) + M_1 + M_2$$
$$(90 \text{ kips} + 120 \text{ kips})x = (120 \text{ kips})(18 \text{ ft}) + 70 \text{ ft-kips}$$
$$+ 70 \text{ ft-kips}$$
$$(210 \text{ kips})x = 2300 \text{ ft-kips}$$
$$x = 11.0 \text{ ft}$$

The length of the footing is

$$L = (\tfrac{1}{2}w_1 + x)(2) = \left(\left(\tfrac{1}{2}\right)(1 \text{ ft}) + 11.0 \text{ ft}\right)(2)$$
$$= 23.0 \text{ ft}$$

The answer is (D).

Why Other Options Are Wrong

(A) This incorrect solution neglects the effects of the applied moments in calculating the location of the resultant load.

(B) This incorrect solution neglects to add one-half the column width when calculating the length of the footing.

(C) This incorrect solution uses factored loads instead of service loads to size the footing. Factored loads are used in the design of the reinforcement for the concrete footing. The size of the footing is based on service (unfactored) loads and soil pressures, because footing design safety is provided by the safety factor in the allowable soil bearing pressure.

56. Statement I is true. Mat foundations are useful in areas where the basement is below the ground water table (GWT) because the mat foundation provides a water barrier, since it is a continuous concrete slab.

Statement II is true. Mat foundations require both top and bottom reinforcement because both positive and negative moments are developed in the foundation.

The answer is (B).

Why Other Options Are Wrong

(A) This incorrect solution only identifies statement I as true and misses the fact that statement II is also true.

(C) This incorrect solution mistakenly identifies the two solutions that are false rather than those that are true. Statements III and IV are false. Mat foundations are suitable where settlements may be a problem or where settlements may be large due, in part, to their rigidity. The rigidity of the mat tends to bridge over areas of erratic soil types or voids.

(D) This incorrect solution wrongly identifies statements III and IV as true. Mat foundations are suitable where settlements may be a problem or where settlements may be large due, in part, to their rigidity. The rigidity of the mat tends to bridge over areas of variable soil types or voids. Statements III and IV are false.

57. The allowable bearing capacity of a single pile is

$$Q_a = \frac{Q_u}{F}$$

The ultimate bearing capacity is the sum of the point-bearing capacity, Q_p, and the skin-friction capacity, Q_f.

$$Q_u = Q_p + Q_f$$
$$Q_p = A_p c N_c$$

From a foundation handbook, for driven piles of virtually all conventional dimensions,

$$N_c = 9$$
$$Q_p = 9 A_p c$$
$$Q_f = A_s f_s$$
$$f_s = c_A + \sigma_h \tan \delta$$

For saturated clay, the angle of external friction, δ, is 0.

$$Q_f = A_s f_s = A_s c_A$$

The area of the pile tip for a steel H-pile is calculated using the area enclosed by the block perimeter. From the AISC ASD manual, Part I, find the block dimensions of an HP12 × 63 to be 11.94 in × 12.125 in.

$$A_p = db_f = \frac{(11.94 \text{ in})(12.125 \text{ in})}{\left(12 \frac{\text{in}}{\text{ft}}\right)^2} = 1.01 \text{ ft}^2$$

$$Q_p = 9A_p c = \frac{(9)(1.01 \text{ ft}^2)\left(400 \frac{\text{lbf}}{\text{ft}^2}\right)}{1000 \frac{\text{lbf}}{\text{kip}}} = 3.6 \text{ kips}$$

In an H-pile, the area between the flanges is assumed to fill with soil that moves with the pile. In calculating the skin area, A_s, of an H-pile, the perimeter is the block perimeter of the pile.

$$\begin{aligned} A_s &= pL \\ &= ((2)(11.94 \text{ in}) + (2)(12.125 \text{ in}))\left(\frac{1 \text{ ft}}{12 \text{ in}}\right)(100 \text{ ft}) \\ &= 401 \text{ ft}^2 \end{aligned}$$

$$Q_f = A_s c_A = \frac{(401 \text{ ft}^2)\left(360 \frac{\text{lbf}}{\text{ft}^2}\right)}{1000 \frac{\text{lbf}}{\text{kip}}} = 144.4 \text{ kips}$$

$$Q_u = Q_p + Q_f = 3.6 \text{ kips} + 144.4 \text{ kips} = 148.0 \text{ kips}$$

With a factor of safety of 3, the allowable load is

$$Q_a = \frac{Q_u}{F} = \frac{148.0 \text{ kips}}{3} = 49.3 \text{ kips} \quad (49 \text{ kips})$$

The answer is (B).

Why Other Options Are Wrong

(A) This incorrect solution does not include the length of the pile in calculating the skin area. The units do not work out.

(C) This solution calculates the perimeter area of the pile incorrectly. When determining the pile capacity, the skin area and the point area of an H-pile should be calculated using the block perimeter and block area of the pile. In this wrong solution, the actual pile perimeter is used, as is the actual area of steel section.

(D) This incorrect solution does not include the factor of safety in determining the allowable capacity.

58. Check that the maximum width of the stiffener, b_t, of 6.0 in does not exceed the limit given by AASHTO Eq. 6.10.11.2.2-1.

$$b_t = 0.48 t \sqrt{\frac{E}{F_{ys}}} \quad [\text{AASHTO Eq. 6.10.11.2.2-1}]$$

$$= (0.48)(0.75 \text{ in})\sqrt{\frac{29{,}000 \frac{\text{kips}}{\text{in}^2}}{50 \frac{\text{kips}}{\text{in}^2}}}$$

$$= 8.67 \text{ in} \quad [\geq 6.0 \text{ in, OK}]$$

From AASHTO Sec. 6.9.4.1, the nominal axial compressive resistance of the bearing stiffener is

$$P_n = 0.66^\lambda F_y A \quad [\text{for } \lambda \leq 2.25, \text{AASHTO Eq. 6.9.4.1-1}]$$

$$P_n = \frac{0.88 F_y A}{\lambda} \quad [\text{for } \lambda > 2.25, \text{AASHTO Eq. 6.9.4.1-2}]$$

Determine the section properties of the bearing stiffener. The effective web width is given in AASHTO Sec. 6.10.11.2.4b as

$$w = 2(9 t_w) = (2)\big((9)(0.5 \text{ in})\big) = 9 \text{ in}$$

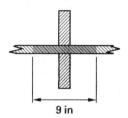

The area is the sum of the area of the bearing stiffener and the area of the beam web.

$$\begin{aligned} A &= 2 b_t t_{\text{stiffener}} + w t_{\text{web}} \\ &= (2)(6 \text{ in})(0.75 \text{ in}) + (9 \text{ in})(0.5 \text{ in}) \\ &= 13.5 \text{ in}^2 \end{aligned}$$

Determine the moment of inertia about the web centerline.

$$\begin{aligned} I &= \frac{bh^3}{12} \\ &= \frac{(0.75 \text{ in})(6.0 \text{ in} + 0.5 \text{ in} + 6.0 \text{ in})^3}{12} \\ &\quad + \frac{(9 \text{ in} - 0.75 \text{ in})(0.5 \text{ in})^3}{12} \\ &= 122.2 \text{ in}^4 \end{aligned}$$

The radius of gyration is

$$r = \sqrt{\frac{I}{A}} = \sqrt{\frac{122 \text{ in}^4}{13.5 \text{ in}^2}} = 3.01 \text{ in}$$

Determine the slenderness ratio. From AASHTO Sec. 6.10.11.2.4a, K equals 0.75.

$$\frac{Kl}{r} = \frac{(0.75)(7.0 \text{ ft})\left(12 \frac{\text{in}}{\text{ft}}\right)}{3.01 \text{ in}} = 20.9$$

$$\lambda = \left(\frac{Kl}{r\pi}\right)^2 \left(\frac{F_y}{E}\right) \quad [\text{AASHTO Eq. 6.9.4.1-3}]$$

$$= \left(\frac{20.9}{\pi}\right)^2 \left(\frac{50 \frac{\text{kips}}{\text{in}^2}}{29{,}000 \frac{\text{kips}}{\text{in}^2}}\right)$$

$$= 0.0763 \quad [\leq 2.25]$$

Therefore, use AASHTO Eq. 6.9.4.1-1.

$$P_n = 0.66^\lambda F_y A = (0.66)^{0.0763}\left(50 \frac{\text{kips}}{\text{in}^2}\right)(13.5 \text{ in}^2)$$

$$= 654 \text{ kips}$$

From AASHTO Sec. 6.5.4.2, ϕ equals 0.9. The factored axial resistance of the effective bearing stiffener column section is

$$P_R = \phi P_n = (0.9)(654 \text{ kips})$$

$$= 589 \text{ kips} \quad (590 \text{ kips})$$

The answer is (B).

Why Other Options Are Wrong

(A) This incorrect solution uses a value of 1.0 for K instead of 0.75.

(C) This incorrect solution finds the nominal resistance instead of the factored resistance.

(D) This incorrect solution mistakenly calculates the factored axial resistance as P_n/ϕ.

59. *MSJC* Sec. 1.18 contains the quality assurance requirements for masonry structures. There are three levels of quality assurance, depending on the type of facility (essential or nonessential) and the method of design.

A residence is a nonessential facility. Since this building was designed using the strength design provisions, it was *not* designed using *MSJC* Chap. 5, Chap. 6, or Chap. 7, which cover empirical design, veneer, and glass unit masonry. *MSJC* Sec. 1.18.2.2 requires such buildings to comply with *MSJC* Table 1.18.2.

MSJC Table 1.18.2 (Level B quality assurance) contains, among others, the following inspection requirements.

Prior to grouting, verify

- grout space
- grade and size of reinforcement
- placement of reinforcement
- proportions of site-prepared grout
- construction of mortar joints

Continuously verify that the placement of grout is in compliance.

Observe preparation of grout specimens, mortar specimens, and/or prisms.

Verify compressive strength of masonry, f'_m, prior to construction.

Statements I, II, and III are true for Level B quality assurance. Statement IV is a provision of Level C quality assurance that requires continual verification of f'_m throughout construction (every 5000 ft^2).

The answer is (C).

Why Other Options Are Wrong

(A) This incorrect solution correctly identifies statements I and II as true, but statement III is also true.

(B) This incorrect solution identifies statements I and III as true, but does not recognize statement II as applicable to Level B quality assurance. The 2008 *MSJC Code* requires continuous verification of grouting.

(D) This incorrect solution mistakenly applies the provisions for buildings designed using *MSJC* Chap. 5, Chap. 6, or Chap. 7. Strength design is found in *MSJC* Chap. 3.

60. According to AISC Table 2-1, the acceptable approaches to assessing stability requirements of a steel building include the direct analysis method, the effective length method, and the first-order analysis method.

The answer is (D).

Why Other Options Are Wrong

(A) This incorrect option is an acceptable method of assessing stability requirements of a steel building.

(B) This incorrect option is an acceptable method of assessing stability requirements of a steel building.

(C) This incorrect option is an acceptable method of assessing stability requirements of a steel building.

Solutions
Lateral Breadth Problems

1. Section 1609 of the IBC covers wind design. This section requires that wind loads be determined using Sec. 6 of ASCE7, which offers three methods of analysis. Method 1 is a simplified method that applies only to buildings. Method 3 is a wind tunnel procedure. For the tank in this case, use the analytical procedure of Method 2 in Sec. 6.5 of ASCE7.

The design wind force for other structures is given in ASCE7 Sec. 6.5.15 as

$$F = q_z G C_f A_f \quad \text{[ASCE7 Eq. 6-28]}$$

The velocity pressure, q_z, is evaluated at the height of the centroid of the projected area, A_f, of the shape. For a 50 ft high tank with a 20 ft diameter, the height of the centroid is

$$z = h - \frac{D}{2} = 50 \text{ ft} - \frac{20 \text{ ft}}{2} = 40 \text{ ft}$$

The velocity pressure is given as

$$q_z = 0.00256 K_z K_{zt} K_d v^2 I \quad \text{[ASCE7 Eq. 6-15]}$$

The velocity pressure coefficient is a function of exposure category. From ASCE7 Sec. 6.5.6.2, determine that a West Coast shoreline exposure is surface roughness D and exposure D [ASCE7 Sec. 6.5.6.3]. Using ASCE7 Table 6-3, determine the velocity pressure coefficient at 40 ft to be

$$K_z = 1.22$$

The topographic factor, K_{zt}, given in ASCE7 Sec. 6.5.7.2 applies to locations that have abrupt changes in general topography. Since the landscape is flat in this case, K_{zt} equals 1.0.

The wind directionality factor for round tanks is given in ASCE7 Table 6-4 as

$$K_d = 0.95$$

The basic wind speed, v, is given as 85 mph. (See also ASCE7 Fig. 6-1.)

The importance factor, I, is given in ASCE7 Table 6-1. For water tanks in building categories III and IV,

$$I = 1.15$$

The velocity pressure is

$$\begin{aligned} q_z &= 0.00256 K_z K_{zt} K_d v^2 I \\ &= (0.00256)(1.22)(1.0)(0.95)\left(85 \ \frac{\text{mi}}{\text{hr}}\right)^2 (1.15) \\ &= 24.7 \ \text{lbf/ft}^2 \end{aligned}$$

The gust effect factor, G, for rigid structures is given in ASCE7 Sec. 6.5.8.1 as 0.85.

The net force coefficients for tanks are found in ASCE7 Fig. 6-21 and are based on the ratio of the structure's height to diameter and on the velocity pressure. In this case,

$$\begin{aligned} h &= 50 \text{ ft} \\ D &= 20 \text{ ft} \\ \frac{h}{D} &= \frac{50 \text{ ft}}{20 \text{ ft}} = 2.5 \\ D\sqrt{q_z} &= (20 \text{ ft})\sqrt{24.7 \ \frac{\text{lbf}}{\text{ft}^2}} = 99.4 \end{aligned}$$

For a moderately smooth tank, interpolate ASCE7 Fig. 6-21 to determine

$$C_f = 0.53$$

The projected area of the circular tank is

$$A_f = \pi\left(\frac{D^2}{4}\right) = \pi\left(\frac{(20 \text{ ft})^2}{4}\right) = 314.2 \ \text{ft}^2$$

The design wind force for the tank is

$$\begin{aligned} F = q_z G C_f A_f &= \left(24.7 \ \frac{\text{lbf}}{\text{ft}^2}\right)(0.85)(0.53)(314.2 \ \text{ft}^2) \\ &= 3496 \ \text{lbf} \quad (3500 \ \text{lbf}) \end{aligned}$$

The answer is (C).

Why Other Options Are Wrong

(A) This incorrect solution determines the velocity pressure (24.7 lbf/ft^2), not the design wind force.

(B) This solution mistakenly identifies the exposure category as C, which applies to terrain that is flat and generally open. Although the terrain at this location is likely flat and open, the correct exposure is D, because the tank is located near a large body of water. The velocity pressure coefficient, K_z, is incorrectly determined to be 1.04.

(D) This incorrect solution reads ASCE7 Table 6-3 for a 50 ft height instead of the height to the centroid of 40 ft and determines K_z to be 1.27.

2. Section 12.14 of ASCE7 contains the simplified design procedure for simple bearing wall buildings. The seismic base shear is

$$V = \left(\frac{FS_{DS}}{R}\right) W \quad \text{[ASCE7 Eq. 12.14-11]}$$

The design elastic response acceleration at the short period, S_{DS}, is given as

$$S_{DS} = \tfrac{2}{3} F_a S_s \quad \text{[ASCE7 Sec. 12.14.8.1]}$$

F_a is 1.4 for soil sites. The maximum considered earthquake ground motion for 0.2 sec spectral response, S_s, is 0.3 g.

$$S_{DS} = \left(\tfrac{2}{3}\right)(1.4)(0.3) = 0.28$$

$F = 1.0$ for one-story buildings. R is the response modification factor from ASCE7 Table 12.14-1 and is a function of the basic seismic-force-resisting system. For a bearing wall system, use Part A of the table. Based on System 9, for ordinary reinforced masonry shear walls, R is 2.0.

$$V = \left(\frac{FS_{DS}}{R}\right) W = \left(\frac{(1.0)(0.28)}{2.0}\right) W$$
$$= 0.14 W$$

The answer is (B).

Why Other Options Are Wrong

(A) This incorrect solution finds the response modification factor in ASCE7 Table 12.14-1 based on System 2, which gives an R of 4 for ordinary reinforced concrete shear walls.

(C) This incorrect solution assumes the design elastic response acceleration at the short period, S_{DS}, to be given as 0.30.

(D) This incorrect solution finds the response modification factor in ASCE7 Table 12.14-1 based on System 2, which gives an R of 4 for ordinary reinforced concrete shear walls, and assumes the design elastic response acceleration at the short period, S_{DS}, to be given as 0.30.

3. Section 12.8.1 of ASCE7 defines seismic base shear as directly proportional to effective seismic weight, W. The effective seismic weight is the total dead load plus portions of other loads as listed in ASCE7. Therefore, as the building dead load increases, so does the seismic base shear. Statement I is true.

ASCE7 states that where the flat snow load exceeds 30 psf, the effective seismic weight shall include 20% of the uniform design snow load. Statement III is also true.

The answer is (B).

Why Other Options Are Wrong

(A) In areas of high seismic activity, buildings perform best when they have a regular floor plan to more uniformly distribute the seismic loads. Irregular plans often have greater building eccentricities and greater differential seismic movements that are harder to accommodate in the building design. Statement II is false.

(C) In the Northridge earthquake and others, buildings with flexible (or soft) first stories and heavy roofs often failed when the flexible first story swayed beyond the design limits and collapsed. In general, it is preferable to have the more flexible story above a stiffer one. Statement IV is false.

(D) Even though statements I and III are true, statement II is false.

4. Section 1613 of the IBC refers to ASCE7 for seismic design, which contains limits on story drift for all seismic design categories. (See ASCE7 Sec. 12.12.) Therefore, statement C is false.

The answer is (C).

Why Other Options Are Wrong

(A) According to the IBC and ASCE7, the design lateral seismic force is directly proportional to the building weight. For example, ASCE7 Sec. 12.14 gives the following equation for seismic base shear.

$$V = \left(\frac{FS_{DS}}{R}\right) W \quad \text{[ASCE7 Eq. 12.14-11]}$$

In this equation, W is the effective seismic weight, which includes the total dead load. As W increases, so does the total design base shear, V. Statement A is true.

(B) Seismic design category is based on seismic use Group and the design spectral response coefficients.

According to IBC Sec. 1604.5, an apartment building is classified as occupancy category II.

Use IBC Table 1613.5.6(1) to determine that S_{DS} is between 0.33 g and 0.50 g, knowing the apartment building is classified as seismic design category C and occupancy category II.

(D) IBC Sec. 1613.5.2 covers site class definitions. The exception in this section states, "When the soil properties are not known in sufficient detail to determine the site class, site class D shall be used…" Statement D is true.

5. The maximum earth pressure, p_{max}, occurs at the base of the wall.

$$p_{max} = K_a \gamma H = (0.36)\left(125 \ \frac{\text{lbf}}{\text{ft}^3}\right)(10 \ \text{ft})$$
$$= 450 \ \text{lbf/ft}^2$$

Draw the load, shear, and moment diagrams for the wall caused by the earth pressure alone, as shown in the *Illustration for Solution 5*.

The roof joists bear on the full width of the concrete wall, so the roof loads are centered on the wall. The load from the storage rack can be assumed to bear at the middle of the 4 in leg of the supporting angle. The eccentricity of the rack load is

$$e = \frac{t_w}{2} + \frac{L_{\text{bearing}}}{2} = \frac{12 \ \text{in}}{2} + \frac{4 \ \text{in}}{2}$$
$$= 8 \ \text{in}$$

The moment due to the eccentricity of the rack load is

$$M = P_{\text{rack}} e = \left(75 \ \frac{\text{lbf}}{\text{ft}}\right)(8 \ \text{in})\left(\frac{1 \ \text{ft}}{12 \ \text{in}}\right)$$
$$= 50 \ \text{ft-lbf/ft}$$

The reactions from the applied loads, including the moment due to the rack load, equal 1872 lbf/ft and 377 lbf/ft at the base and top of the wall, respectively.

The shear at a distance x greater than or equal to 10 ft from the top of the wall is calculated as

$$V_x = 377 \ \frac{\text{lbf}}{\text{ft}} - \left(\left(45 \ \frac{\text{lbf}}{\text{ft}^3}\right)(x - 10 \ \text{ft})\right)\left(\tfrac{1}{2}\right)(x - 10 \ \text{ft})$$

The maximum moment occurs at 14.09 ft from the top of the wall, where shear, V, equals 0 lbf. The moment at this point is

$$M_{max} = \left(377 \ \frac{\text{lbf}}{\text{ft}}\right)(14.09 \ \text{ft})$$
$$- \left(\left(\tfrac{1}{2}\right)(4.09 \ \text{ft})\left(45 \ \frac{\text{lbf}}{\text{ft}^3}\right)(4.09 \ \text{ft})\right)$$
$$\times \left(\tfrac{1}{3}\right)(4.09 \ \text{ft}) - 50 \ \frac{\text{ft-lbf}}{\text{ft}}$$
$$= 4749 \ \text{ft-lbf/ft} \quad (4750 \ \text{ft-lbf/ft})$$

The answer is (C).

Why Other Options Are Wrong

(A) This incorrect solution mistakenly includes the roof load in the calculation of moment due to the eccentric loads.

(B) This incorrect solution does not convert the units for the eccentricity from inches to feet when calculating the moment due to the eccentric load.

(D) This incorrect solution adds the maximum moment caused by the eccentricity of the rack load to the maximum moment caused by the earth pressure.

Illustration for Solution 5

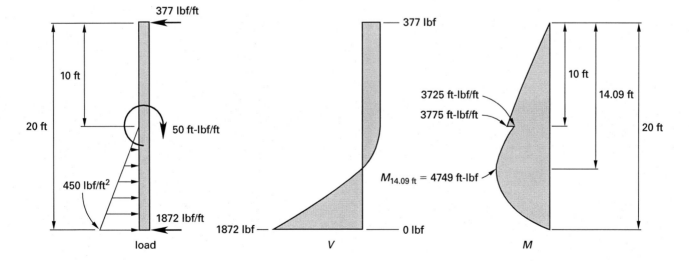

6. Find the rigidity of the walls. Rigidity is the relative stiffness of the walls and is proportional to the inverse of the shear and flexural deflection of the walls. However, for squat walls with h/L no more than 0.25, it is reasonably accurate to calculate the rigidity based on shear alone. For walls with h/L greater than 0.25 and less than 4, both flexure and shear must be considered. For very tall walls, the contribution from shear deformations is very small, and the rigidity can be based on flexure alone.

Since walls B, C, and E have the shortest length, the critical aspect ratio for these walls is

$$\frac{h}{L} = \frac{9 \text{ ft}}{36 \text{ ft}} = 0.25$$

Therefore, the rigidity can be based on shear alone. Assume the walls all have the same modulus of elasticity. Since the walls are all the same height, h_j, the rigidity is proportional to the area of the walls. If the walls had the same thickness, the shear rigidity would be proportional to wall length alone.

$$r_j = \frac{k_j}{\sum k_i}$$

$$k_j = \frac{A_j E_j}{h_j}$$

$$r_j = \frac{k_j}{k_j + k_k} = \frac{A_j E_j}{h_j \left(\frac{A_j E_j}{h_j} + \frac{A_k E_k}{h_k}\right)}$$

$$r_j \propto A = tL$$

$$r_A \propto (16 \text{ in})\left(\frac{1 \text{ ft}}{12 \text{ in}}\right)(60 \text{ ft}) = 80 \text{ ft}^2$$

$$r_B \propto (16 \text{ in})\left(\frac{1 \text{ ft}}{12 \text{ in}}\right)(36 \text{ ft}) = 48 \text{ ft}^2$$

$$r_C \propto (12 \text{ in})\left(\frac{1 \text{ ft}}{12 \text{ in}}\right)(40 \text{ ft}) = 40 \text{ ft}^2$$

$$r_D \propto (12 \text{ in})\left(\frac{1 \text{ ft}}{12 \text{ in}}\right)(36 \text{ ft}) = 36 \text{ ft}^2$$

$$r_E \propto (16 \text{ in})\left(\frac{1 \text{ ft}}{12 \text{ in}}\right)(36 \text{ ft}) = 48 \text{ ft}^2$$

Find the center of rigidity in each direction.

The distance to the center of rigidity from the south end is

$$\bar{y} = \frac{\sum r_i y_i}{\sum r_i} = \frac{r_A y_A + r_D y_D}{r_A + r_D}$$

$$= \frac{(80 \text{ ft}^2)(160 \text{ ft}) + (36 \text{ ft}^2)(40 \text{ ft})}{80 \text{ ft}^2 + 36 \text{ ft}^2}$$

$$= 123 \text{ ft}$$

The distance to the center of rigidity from the west end is

$$\bar{x} = \frac{\sum r_i x_i}{\sum r_i} = \frac{r_B x_B + r_C x_C + r_E x_E}{r_B + r_C + r_E}$$

$$= \frac{(48 \text{ ft}^2)(80 \text{ ft}) + (40 \text{ ft}^2)(50 \text{ ft}) + (48 \text{ ft}^2)(20 \text{ ft})}{48 \text{ ft}^2 + 40 \text{ ft}^2 + 48 \text{ ft}^2}$$

$$= 50.0 \text{ ft}$$

The answer is (D).

Why Other Options Are Wrong

(A) This incorrect solution does not include the effects of the different wall thicknesses when calculating the rigidities.

(B) This incorrect solution does not include the effects of the different wall thicknesses and makes a calculation error when determining the distance to the center of rigidity from the west end by calculating $\sum r_i$ equal to 102 ft instead of 112 ft.

(C) This incorrect solution correctly finds the center of rigidity but reverses the direction from which the distance is measured when choosing the answer.

7. A roof framed with steel joists and corrugated steel deck is typically considered a flexible diaphragm because of its light weight and flexibility. A rigid diaphragm can be achieved if 2–3 in of concrete are used on the steel decking.

For a flexible diaphragm, the lateral loads are distributed according to tributary area. If the roof were a rigid diaphragm (i.e., concrete), the lateral loads would be distributed according to rigidity.

In this case, the shear load can be distributed according to the tributary width, w.

$$V_A = \left(\frac{w_A}{\sum w_i}\right) V_{total} = \left(\frac{15 \text{ ft} + \left(\frac{1}{2}\right)(20 \text{ ft})}{85 \text{ ft}}\right)(600 \text{ kips})$$

$$= 176 \text{ kips}$$

The answer is (B).

Why Other Options Are Wrong

(A) This incorrect solution calculates the tributary width of wall A using one-half the distance to the edge of the building instead of the full tributary width.

(C) This incorrect solution distributes the lateral load based on relative rigidities. Relative rigidities can only be used if the diaphragm is rigid. In addition, when loads are based on rigidities, eccentricity of the load (torsion) must also be considered.

(D) This incorrect solution calculates the tributary width of wall A using the full distance between walls instead of one-half the distance between the walls.

8. The shear force in the walls is a combination of the total shear force on the building and the shear induced by the torsional moment. If the resultant wind load, W, does not pass through the center of rigidity (c.r.) of the structure, the resultant creates a torsional moment that is resisted by shear in the walls.

$$V = \frac{Wr_i}{\sum r_i} + \frac{M_t x r_i}{I_p}$$

$$W = wL = \left(200 \; \frac{\text{lbf}}{\text{ft}}\right)(160 \text{ ft})$$

$$= 32{,}000 \text{ lbf} \quad (32 \text{ kips})$$

$$I_p = I_{xx} + I_{yy}$$

$$I_{xx} = \sum Ay^2$$

$$I_{yy} = \sum Ax^2$$

Calculate I_{xx} using walls B and C. Calculate I_{yy} using walls A and D. x and y are the distances from the centroids of each wall area to the center of rigidity of the building.

$$\begin{aligned}
I_{xx} &= \sum Ay^2 \\
&= (20 \text{ ft})(1 \text{ ft})(10 \text{ ft})^2 + (20 \text{ ft})(1 \text{ ft})(10 \text{ ft})^2 \\
&= 4000 \text{ ft}^4 \\
I_{yy} &= \sum Ax^2 \\
&= (50 \text{ ft})(1 \text{ ft})(98.3 \text{ ft} - 65 \text{ ft})^2 \\
&\quad + (40 \text{ ft})(1 \text{ ft})(140 \text{ ft} - 98.3 \text{ ft})^2 \\
&= 125{,}000 \text{ ft}^4 \\
I_p &= I_{xx} + I_{yy} = 4000 \text{ ft}^4 + 125{,}000 \text{ ft}^4 \\
&= 129{,}000 \text{ ft}^4
\end{aligned}$$

The torsional moment is the product of the resultant and the distance between the centroid of the load and the center of rigidity.

$$\begin{aligned}
M_t &= W\overline{x} \\
&= (32 \text{ kips})(98.3 \text{ ft} - 80 \text{ ft}) \\
&= 586 \text{ ft-kips}
\end{aligned}$$

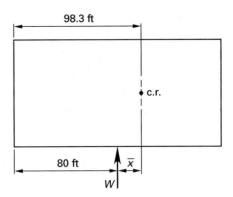

Rigidity is the relative stiffness of the walls. In this case, all walls are 12 in thick. Assume the walls all have the same modulus of elasticity and height, L_j.

$$r_j = \frac{k_j}{\sum k_i}$$

$$k_j = \frac{A_j E_j}{L_j}$$

$$r_j = \frac{k_j}{k_j + k_k} = \frac{\dfrac{A_j E_j}{L_j}}{\left(\dfrac{A_j E_j}{L_j} + \dfrac{A_k E_k}{L_k}\right)}$$

Since E_j and E_k are equal and L_j and L_k are equal, the rigidity becomes

$$r_j = \frac{A_j}{A_j + A_k}$$

For wall A,

$$\begin{aligned}
V_a &= \frac{Wr_a}{\sum r_i} + \frac{M_t x r_a}{I_p} \\
&= \frac{(32 \text{ kips})(50 \text{ ft}^2)}{50 \text{ ft}^2 + 40 \text{ ft}^2} \\
&\quad + \frac{(586 \text{ ft-kips})(98.3 \text{ ft} - 65 \text{ ft})(50 \text{ ft}^2)}{129{,}000 \text{ ft}^4} \\
&= 25.3 \text{ kips} \quad (25 \text{ kips})
\end{aligned}$$

The answer is (D).

Why Other Options Are Wrong

(A) This incorrect solution finds the shear load on wall D instead of wall A.

(B) This incorrect solution includes the east and west walls in calculating rigidity and does not include the effects of the eccentricity of the resultant (torsional moment).

For wall A,

$$V_a = \frac{Wr_a}{\sum r_i} = \frac{(32 \text{ kips})(50 \text{ ft}^2)}{50 \text{ ft}^2 + 20 \text{ ft}^2 + 20 \text{ ft}^2 + 40 \text{ ft}^2}$$
$$= 12.3 \text{ kips} \quad (12 \text{ kips})$$

(C) This incorrect solution does not include the effects of the torsional moment.

9. The load combinations to be considered are given in IBC Sec. 1605.3 for allowable stress design. For this case, consider

$$D$$
$$D + (L_{\text{roof}} \text{ or } R)$$
$$D + 0.75(L_{\text{roof}} \text{ or } R)$$
$$D + W$$
$$D + 0.75W + 0.75(L_{\text{roof}} \text{ or } R)$$
$$0.6D + W$$

Calculate the dead load on the roof deck.

$$\text{joists} = 5 \text{ lbf/ft}^2$$
$$\text{roof deck} = 3 \text{ lbf/ft}^2$$
$$\text{rigid insulation} = 3 \text{ lbf/ft}^2$$
$$\text{felt and gravel} = 5 \text{ lbf/ft}^2$$
$$\text{total} = 16 \text{ lbf/ft}^2$$

The live load for the roof deck is 20 lbf/ft^2, and the rain load for the roof is 10 lbf/ft^2.

Determine the wind load using IBC Sec. 1609 and Chap. 6 of ASCE7. The building meets the requirements of the simplified provisions for low-rise buildings and can be designed in accordance with ASCE7 Sec. 6.4.

The basic wind speed, v, is given as 100 mph.

From ASCE7 Table 1-1, determine that a warehouse is building category II. From ASCE7 Table 6-1, the importance factor, I, is 1.00.

From IBC Sec. 1609.4, determine the surface roughness and the exposure categories to be B. From ASCE7 Fig. 6-2, determine the height and exposure adjustment coefficient, λ, for a 30 ft roof height and exposure B to be 1.00.

The simplified design wind pressure for the MWFRS is

$$p_s = \lambda K_{zt} I p_{s30} \quad [\text{ASCE7 Eq. 6-1}]$$

From ASCE7 Fig. 6-2, determine the roof pressures to be zones E, F, G, and H. These zones account for localized higher pressures.

From ASCE7 Sec. 6.5.7, determine the topographic factor K_{zt} to be 1.00. Using ASCE7 Fig. 6-2, determine the maximum and minimum simplified pressures, p_{s30}, for a 100 mph wind speed and a flat roof for zones E, F, G, and H. The maximum and minimum pressures are for zones E and H, respectively.

$$p_{s30,\text{max}} = -19.1 \text{ lbf/ft}^2$$
$$p_{s30,\text{min}} = -8.4 \text{ lbf/ft}^2$$

Calculate the maximum and minimum wind pressures.

$$p_s = \lambda K_{zt} I p_{s30}$$
$$p_{s,\text{max}} = (1.00)(1.00)(1.00)\left(-19.1 \frac{\text{lbf}}{\text{ft}^2}\right) = -19.1 \text{ lbf/ft}^2$$
$$p_{s,\text{min}} = (1.00)(1.00)(1.00)\left(-8.4 \frac{\text{lbf}}{\text{ft}^2}\right) = -8.4 \text{ lbf/ft}^2$$

Determine the worst-case load.

By inspection, determine that dead load alone is not the critical combination and that the roof rain load is less than the roof live load.

$$D + L_{\text{roof}} = 16 \frac{\text{lbf}}{\text{ft}^2} + 20 \frac{\text{lbf}}{\text{ft}^2} = 36 \text{ lbf/ft}^2$$

$D + W$:

$$D + p_{s,\text{max}} = 16 \frac{\text{lbf}}{\text{ft}^2} - 19.1 \frac{\text{lbf}}{\text{ft}^2} = -3.1 \text{ lbf/ft}^2$$
$$D + p_{s,\text{min}} = 16 \frac{\text{lbf}}{\text{ft}^2} - 8.4 \frac{\text{lbf}}{\text{ft}^2} = 7.6 \text{ lbf/ft}^2$$

$D + 0.75W + 0.75L_{\text{roof}}$:

$$D + 0.75(p_{s,\text{max}} + L_{\text{roof}}) = 16 \frac{\text{lbf}}{\text{ft}^2} + (0.75)$$
$$\times \left(-19.1 \frac{\text{lbf}}{\text{ft}^2} + 20 \frac{\text{lbf}}{\text{ft}^2}\right)$$
$$= 17 \text{ lbf/ft}^2$$

$$D + 0.75(p_{s,\text{min}} + L_{\text{roof}}) = 16 \frac{\text{lbf}}{\text{ft}^2} + (0.75)$$
$$\times \left(-8.4 \frac{\text{lbf}}{\text{ft}^2} + 20 \frac{\text{lbf}}{\text{ft}^2}\right)$$
$$= 25 \text{ lbf/ft}^2$$

$0.6D + W$:

$$0.6D + p_{s,\text{max}} = (0.6)\left(16\ \frac{\text{lbf}}{\text{ft}^2}\right) - 19.1\ \frac{\text{lbf}}{\text{ft}^2} = -9.5\ \text{lbf/ft}^2$$

$$0.6D + p_{s,\text{min}} = (0.6)\left(16\ \frac{\text{lbf}}{\text{ft}^2}\right) - 8.4\ \frac{\text{lbf}}{\text{ft}^2} = 1.2\ \text{lbf/ft}^2$$

The critical load case is that which is maximum, $D + L_{\text{roof}}$. The critical design load is 36 lbf/ft^2.

The answer is (C).

Why Other Options Are Wrong

(A) This incorrect solution gives the largest uplift load from the $0.6D + W$ load combination (-9.5 lbf/ft^2).

(B) This incorrect solution gives the maximum uplift pressure (-19 lbf/ft^2) and does not consider the load combinations.

(D) This incorrect solution uses the load combinations for strength design instead of those for allowable stress.

The critical design case is $1.2D + 1.6L_{\text{roof}}$.

10. ASCE7 Sec. 6.4 covers method 1—simplified procedure for wind design. Use ASCE7 Sec. 6.4.2.2 to determine the net design wind pressures for components and cladding.

ASCE7 Fig. 6-3 indicates the pressure zones based on the geometry of the building. For a building with a gable roof whose slope is greater than 7° but no more than 45°, the area at the wall corner is classified as zone 5. From the table in Fig. 6-3, for a wall in zone 5, an effective wind area of 20 ft^2, and a basic wind speed of 100 mph, the net design wind pressures are

$$p_{\text{net 30}} = 17.2\ \text{lbf/ft}^2$$
$$p_{\text{net 30}} = -22.5\ \text{lbf/ft}^2$$

The adjustment factor for building height and exposure, λ, is determined from ASCE7 Fig. 6-2 from the mean roof height and exposure category. For a mean roof height of 40 ft and exposure B, λ is 1.09.

Using ASCE7 Eq. 6-2, the net design wind pressure on the cladding is

$$p_{\text{net}} = \lambda K_{zt} I p_{\text{net 30}} = (1.09)(1.0)(1.0)\left(17.2\ \frac{\text{lbf}}{\text{ft}^2}\right)$$
$$= 18.7\ \text{lbf/ft}^2\quad (19\ \text{lbf/ft}^2)$$

$$p_{\text{net}} = \lambda K_{zt} I p_{\text{net 30}} = (1.09)(1.0)(1.0)\left(-22.5\ \frac{\text{lbf}}{\text{ft}^2}\right)$$
$$= -24.5\ \text{lbf/ft}^2\quad (-25\ \text{lbf/ft}^2)$$

The answer is (D).

Why Other Options Are Wrong

(A) This incorrect solution finds the net design pressures in ASCE7 Fig. 6-3 for zone 4 walls, rather than zone 5, and neglects to apply the adjustment factor to the net design wind pressures.

(B) This incorrect solution neglects to apply the adjustment factor to the net design wind pressures.

(C) This incorrect solution finds the net design wind pressures in ASCE7 Fig. 6-3 for zone 4 walls, rather than zone 5.

11. The portal method assumes that exterior columns carry half the shear of interior columns and that the interior columns carry equal shear.

$$V_{\text{I}} = 0.5 V_{\text{II}}$$
$$V_{\text{IV}} = 0.5 V_{\text{II}}$$
$$V_{\text{II}} = V_{\text{III}}$$

Thus,

$$F = V_{\text{I}} + V_{\text{II}} + V_{\text{III}} + V_{\text{IV}}$$
$$= 0.5 V_{\text{II}} + V_{\text{II}} + V_{\text{II}} + 0.5 V_{\text{II}}$$
$$= 3 V_{\text{II}}$$
$$V_{\text{II}} = \tfrac{1}{3}F$$

The answer is (D).

Why Other Options Are Wrong

(A) This incorrect solution divides the shear in each column in half and assumes all columns carry equal shear.

(B) This incorrect solution assumes the interior columns carry half the shear of exterior columns.

(C) This incorrect solution assumes all columns carry equal shear.

12. The chords at lines 1 and 2 resist bending due to the diaphragm loads when the loading is in the direction shown. At both lines, the maximum chord forces are equal and occur at mid-span where the maximum moment develops.

The answer is (A).

Why Other Options Are Wrong

(B) This incorrect solution assumes the two chords in tension have different maximum forces.

(C) This incorrect solution assumes the chord force opposite the load can be neglected.

(D) This incorrect solution calculates the shear force instead of the maximum chord force.

13. The chord force is the bending moment per unit depth of the diaphragm. The shear load at the parallel walls is

$$V = \frac{wL}{2} = \frac{\left(500\ \frac{\text{lbf}}{\text{ft}}\right)(80\ \text{ft})}{2}$$
$$= 20{,}000\ \text{lbf}$$

The shear and bending moment diaphragms across the length of the chord are as shown.

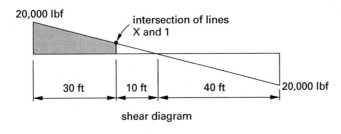

shear diagram

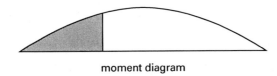

moment diagram

$$\frac{20{,}000\ \text{lbf}}{40} = \frac{V}{10}$$
$$V = 5000\ \text{lbf} \quad [\text{at lines X and 1}]$$

The moment at line 1 is calculated as the area under the shear diagram.

The chord force is the bending moment divided by the diaphragm depth.

$$M = (5000\ \text{lbf})(30\ \text{ft}) + \frac{(15{,}000\ \text{lbf})(30\ \text{ft})}{2}$$
$$= 375{,}000\ \text{ft-lbf}$$
$$C = \frac{M}{b} = \frac{375{,}000\ \text{ft-lbf}}{40\ \text{ft}}$$
$$= 9375\ \text{lbf} \quad (9400\ \text{lbf})$$

The answer is (C).

Why Other Options Are Wrong

(A) This incorrect solution divides the bending moment by the diaphragm length instead of the diaphragm depth.

(B) This incorrect solution uses a distance of 40 ft instead of 30 ft for the distance of the intersection of lines X and 1 to the west wall and divides the bending moment by the diaphragm length instead of the diaphragm depth.

(D) This incorrect solution uses a distance of 40 ft instead of 30 ft for the distance of the intersection of lines X and 1 to the west wall.

14. Section 11.9 of ACI 318 covers design of shear walls. Since the wall supports a uniform load, w_u, that is positive, the wall is in compression and the concrete shear strength is

$$V_c = 2\sqrt{f'_c}hd \quad [\text{ACI 318 Sec. 11.9.5}]$$
$$= 2\sqrt{4000\ \frac{\text{lbf}}{\text{in}^2}}\,(12\ \text{in})(110\ \text{in})\left(\frac{1\ \text{kip}}{1000\ \text{lbf}}\right)$$
$$= 167\ \text{kips}$$
$$V_n = V_c + V_s = 167\ \text{kips} + 700\ \text{kips}$$
$$= 867\ \text{kips}$$

The nominal shear strength of the wall is limited by ACI 318 Sec. 11.9.3 to

$$V_n = 10\sqrt{f'_c}hd$$
$$= 10\sqrt{4000\ \frac{\text{lbf}}{\text{in}^2}}\,(12\ \text{in})(110\ \text{in})\left(\frac{1\ \text{kip}}{1000\ \text{lbf}}\right)$$
$$= 834.8\ \text{kips} \quad [\text{controls}]$$

Therefore, the nominal shear strength of the wall is 834.8 kips (830 kips).

The answer is (C).

Why Other Options Are Wrong

(A) This incorrect solution calculates the shear strength of the concrete instead of the shear strength of the wall.

(B) This incorrect solution calculates the factored shear strength of the wall.

(D) This incorrect solution ignores the limit on nominal shear strength found in ACI 318 Sec. 11.10.3.

15. From ACI 318 Sec. 11.9.9, the amount of horizontal shear reinforcement required depends upon whether or not $V_u < \phi V_c/2$.

$$\phi V_c = \phi 2\sqrt{f'_c}hd$$
$$\phi = 0.75$$
$$h = 10\ \text{in}$$
$$d = 110\ \text{in}$$

$$\phi V_c = \phi 2\sqrt{f'_c}hd$$
$$= (0.75)(2)\sqrt{6000 \ \frac{\text{lbf}}{\text{in}^2}}(10 \text{ in})(110 \text{ in})\left(\frac{1 \text{ kip}}{1000 \text{ lbf}}\right)$$
$$= 128 \text{ kips}$$
$$\frac{\phi V_c}{2} = \frac{128 \text{ kips}}{2}$$
$$= 64.0 \text{ kips} \quad [< V_u = 110 \text{ kips}]$$

Therefore, the reinforcement must be designed in accordance with ACI 318 Sec. 11.9. Since $\phi V_c > V_u$, the area of reinforcement is controlled by the minimum area requirements.

ACI 318 Sec. 11.9.9.2 states that the ratio of horizontal shear reinforcement area to the gross concrete area of the vertical section shall not be less than 0.0025.

$$\rho_t = \frac{A_v}{A_g} = 0.0025$$
$$A_v = 0.0025 A_g = 0.0025 h h_w$$
$$= (0.0025)(10 \text{ in})(30 \text{ ft})\left(12 \ \frac{\text{in}}{\text{ft}}\right)$$
$$= 9.00 \text{ in}^2$$
$$\frac{A_v}{h_w} = \frac{9.00 \text{ in}^2}{30 \text{ ft}} = 0.30 \text{ in}^2/\text{ft}$$

No. 5 bars at 12 in spacing provide 0.31 in^2/ft.

Check the spacing limits in ACI 318 Sec. 11.9.9.5. The maximum spacing is the smallest of

$$s_2 \leq \frac{l_w}{3} = \frac{120 \text{ in}}{3} = 40 \text{ in}$$
$$s_2 \leq 3h = (3)(10 \text{ in}) = 30 \text{ in}$$
$$s_2 \leq 18 \text{ in} \quad [\text{controls}]$$

No. 5 bars at 12 in is OK.

The answer is (C).

Why Other Options Are Wrong

(A) This incorrect solution uses the horizontal cross section, instead of the vertical section, in calculating A_g.

(B) This incorrect solution calculates V_s as

$$V_s = \phi V_c - V_u = 128 \text{ kips} - 110 \text{ kips}$$
$$= 18 \text{ kips}$$

Then, it sizes the reinforcement accordingly.

(D) This incorrect solution uses the minimum horizontal reinforcement requirements found in ACI 318 Chap. 14. ACI 318 Chap. 14 can only be used when $V_u < \phi V_c/2$.

16. The ratio of vertical shear reinforcement area to gross concrete area of a horizontal section for a shear wall is given in ACI 318 Sec. 11.9.9.4 as the larger of 0.0025 and

$$\rho_l = 0.0025 + (0.5)\left(2.5 - \frac{h_w}{l_w}\right)(\rho_t - 0.0025)$$

[ACI 318 Eq. 11-32]

This value need not exceed the required horizontal shear reinforcement, ρ_t.

$$h_w = (10 \text{ ft})\left(12 \ \frac{\text{in}}{\text{ft}}\right) = 120 \text{ in}$$
$$l_w = (30 \text{ ft})\left(12 \ \frac{\text{in}}{\text{ft}}\right) = 360 \text{ in}$$
$$\rho_l = 0.0025 + (0.5)\left(2.5 - \frac{h_w}{l_w}\right)(\rho_t - 0.0025)$$
$$= 0.0025 + (0.5)\left(2.5 - \frac{120 \text{ in}}{360 \text{ in}}\right)(0.0040 - 0.0025)$$
$$= 0.00413$$
$$\rho_t = 0.0040$$

Since ρ_l must be less than or equal to ρ_t, ρ_l is 0.0040.

The answer is (C).

Why Other Options Are Wrong

(A) This incorrect solution reverses the height and length of the wall in ACI 318 Eq. 11-32 and ignores the minimum reinforcement requirement of ACI 318 Sec. 11.9.9.4.

(B) This incorrect solution reverses the height and length of the wall in ACI 318 Eq. 11-32.

(D) This incorrect solution correctly calculates the vertical shear reinforcement ratio but does not apply the limit of $\rho_l \leq \rho_t$ given in ACI 318 Sec. 11.9.9.4.

17. The beams framing the first-floor corner column are exterior beams and have flanges on only one side. Section 13.2.4 of ACI 318 defines a beam in a two-way slab system as including that portion of the slab on each side of the beam, b_f, extending a distance equal to the projection of the beam above or below the slab, whichever is greater, but not greater than four times the slab thickness.

$$b_f = 14 \text{ in} \leq 4t$$
$$= (4)(6 \text{ in})$$
$$= 24 \text{ in}$$

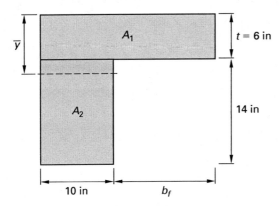

Compute I_b.

	A (in²)	y (in)	Ay (in³)	$A(y-\bar{y})^2$ (in⁴)	I_o (in⁴)
A_1	144	3	432	3500	432
A_2	140	13	1820	3599	2287
totals	284		2252	7099	2719

$$\bar{y} = \frac{\sum Ay}{\sum A} = \frac{2252 \text{ in}^3}{284 \text{ in}^2} = 7.93 \text{ in}$$

$$I_b = I_o + A(y-\bar{y})^2 = 2719 \text{ in}^4 + 7099 \text{ in}^4$$
$$= 9818 \text{ in}^4$$

According to ACI 318 Sec. R10.10.4.1, when calculating Ψ for flexural members,

$$I_{b,e} = 0.35 I_b = (0.35)(9818 \text{ in}^4)$$
$$= 3436 \text{ in}^4 \quad (3400 \text{ in}^4)$$

The answer is (A).

Why Other Options Are Wrong

(B) This incorrect solution uses the definition of a T-beam found in ACI 318 Sec. 8.10.3 when determining the extent beam flange. This approach is applicable to one-way slabs only.

(C) This incorrect solution calculates the beam flange correctly but does not recognize the beam as being an exterior beam with a flange on only one side.

(D) In this incorrect solution, the gross moment of inertia is calculated instead of I_e.

18. Use the following equation for calculating the maximum biaxial load.

$$\frac{1}{P_{\text{biaxial}}} = \frac{1}{P_x} + \frac{1}{P_y} - \frac{1}{P_a}$$

Use column interaction diagrams (such as those in the *Masonry Designers' Guide* or a similar reference) to determine the maximum axial load in each direction.

In the x-direction,

$$e_x = 5.0 \text{ in}$$

The section properties of the column are based on the actual column dimensions. For a 20 in brick masonry column, the actual dimensions are

$$b = t = 19.625 \text{ in}$$
$$A_s = \text{four no. 6 bars} = (4)(0.44 \text{ in}^2) = 1.76 \text{ in}^2$$

$$\rho_t = \frac{A_s}{bt} = \frac{1.76 \text{ in}^2}{(19.625 \text{ in})(19.625 \text{ in})} = 0.0046$$

$$E_m = 700 f'_m \quad \text{[for clay masonry]}$$
$$= (700)\left(3500 \frac{\text{lbf}}{\text{in}^2}\right)$$
$$= 2.45 \times 10^6 \text{ lbf/in}^2$$
$$E_s = 29 \times 10^6 \text{ lbf/in}^2$$

$$n = \frac{E_s}{E_m} = \frac{29 \times 10^6 \frac{\text{lbf}}{\text{in}^2}}{2.45 \times 10^6 \frac{\text{lbf}}{\text{in}^2}} = 11.8$$

$$n\rho_t = (11.8)(0.0046) = 0.054 \quad [\text{Use } 0.05.]$$

$$\frac{e_x}{t} = \frac{5.0 \text{ in}}{19.625 \text{ in}} = 0.255$$

From column interaction diagrams, $P/F_b bt$ is 0.38.

$$P_x = 0.38 F_b bt = (0.38)\left(\tfrac{1}{3}\right) f'_m bt$$
$$= (0.38)\left(\tfrac{1}{3}\right)\left(3500 \frac{\text{lbf}}{\text{in}^2}\right)(19.625 \text{ in})(19.625 \text{ in})$$
$$= 170{,}746 \text{ lbf}$$

In the y-direction, e is 6.2 in.

$$\frac{e_y}{t} = \frac{6.2 \text{ in}}{19.625 \text{ in}} = 0.316$$
$$n\rho_t = (11.8)(0.0046) = 0.054 \quad [\text{Use } 0.05.]$$

From column interaction diagrams, $P/F_b bt$ is 0.26.

$$P_y = 0.26 F_b bt = (0.26)\left(\tfrac{1}{3}\right) f'_m bt$$
$$= (0.26)\left(\tfrac{1}{3}\right)\left(3500 \frac{\text{lbf}}{\text{in}^2}\right)(19.625 \text{ in})(19.625 \text{ in})$$
$$= 116{,}826 \text{ lbf}$$

Find P_a. Since $h/r < 99$, use *Building Code Requirements for Masonry Structures* (MSJC) Eq. 2-20. The contribution from the steel is relatively small. Ignoring the steel,

$$P_a = (0.25 f'_m A_n)\left(1 - \left(\frac{h}{140r}\right)^2\right)$$

$$= \begin{pmatrix} (0.25)\left(3500 \; \dfrac{\text{lbf}}{\text{in}^2}\right)(19.625 \text{ in}) \\ \times (19.625 \text{ in}) \end{pmatrix} \left(1 - \left(\frac{72}{140}\right)^2\right)$$

$$= 247{,}866 \text{ lbf}$$

$$\frac{1}{P_{\text{biaxial}}} = \frac{1}{P_x} + \frac{1}{P_y} - \frac{1}{P_a}$$

$$= \frac{1}{170{,}746 \text{ lbf}} + \frac{1}{116{,}826 \text{ lbf}} - \frac{1}{247{,}866 \text{ lbf}}$$

$$P_{\text{biaxial}} = 96{,}321 \text{ lbf} \quad (96{,}000 \text{ lbf})$$

The answer is (B).

Why Other Options Are Wrong

(A) This incorrect solution makes a calculation error in the biaxial equation, adding the inverse of the pure axial load rather than subtracting it.

(C) This incorrect solution uses the diameter of a no. 6 bar as its area when calculating A_s.

(D) This incorrect solution calculates the maximum axial load for zero eccentricity using *MSJC* Eq. 2-20.

19. Calculate the number of bolts per joist.

$$n = \frac{\text{joist spacing}}{\text{bolt spacing}} = \frac{6.0 \text{ ft}}{(16 \text{ in})\left(\dfrac{1 \text{ ft}}{12 \text{ in}}\right)}$$

$$= 4.5 \text{ bolts/joist}$$

Calculate the loads per bolt.

The applied shear load per bolt is

$$b_v = \frac{R}{n} = \frac{3700 \; \dfrac{\text{lbf}}{\text{joist}}}{4.5 \; \dfrac{\text{bolts}}{\text{joist}}}$$

$$= 822 \text{ lbf/bolt}$$

The prying tension per joist is

$$T = \frac{Rx}{y} = \frac{\left(3700 \; \dfrac{\text{lbf}}{\text{joist}}\right)(2.5 \text{ in})}{3 \text{ in}}$$

$$= 3083 \text{ lbf/joist}$$

The prying tension per bolt is

$$b_a = \frac{T}{n} = \frac{3083 \; \dfrac{\text{lbf}}{\text{joist}}}{4.5 \; \dfrac{\text{bolts}}{\text{joist}}}$$

$$= 685 \text{ lbf/bolt}$$

Determine the allowable loads using *Building Code Requirements for Masonry Structures* (*MSJC*). The allowable load in shear, B_v, is the smallest of the allowable shear loads governed by masonry breakout [*MSJC* Eq. 2-6], by masonry crushing [*MSJC* Eq. 2-7], by anchor bolt pryout [*MSJC* Eq. 2-8], and by steel yielding [*MSJC* Eq. 2-9]. $A_{p,v}$ is the projected area for shear and is calculated from the anchor bolt edge distance in the direction of the load, l_{be}, and *MSJC* Eq. 1-3. Since the anchor bolt is not near the top or bottom of the wall, $A_{p,v}$ will not control. Therefore, $B_{v,b}$ can be disregarded.

First, calculate the projected area for axial tension, $A_{p,t}$. The effective embedment length, l_b, is approximately 6 in.

$$A_{p,t} = \pi l_b^2$$

$$= \pi (6 \text{ in})^2$$

$$= 113 \text{ in}^2 \quad [\textit{MSJC} \text{ Eq. 1-2}]$$

Calculate the allowable shear loads. Try a $^3/_4$ in diameter bolt, which has an area, A_b, of 0.44 in^2.

$$B_{v,c} = 350\sqrt[4]{f'_m A_b} \quad [\textit{MSJC} \text{ Eq. 2-7}]$$

$$= 350\sqrt[4]{\left(1500 \; \dfrac{\text{lbf}}{\text{in}^2}\right)(0.44 \text{ in}^2)}$$

$$= 1774 \text{ lbf} \quad [\text{controls}]$$

$$B_{v,\text{pry}} = 2.5 A_{p,t}\sqrt{f'_m} \quad [\textit{MSJC} \text{ Eq. 2-8}]$$

$$= (2.5)(113 \text{ in}^2)\sqrt{1500 \; \dfrac{\text{lbf}}{\text{in}^2}}$$

$$= 10{,}941 \text{ lbf}$$

$$B_{v,s} = 0.36 A_b f_y \quad [\textit{MSJC} \text{ Eq. 2-9}]$$

$$= (0.36)(0.44 \text{ in}^2)\left(36{,}000 \; \dfrac{\text{lbf}}{\text{in}^2}\right)$$

$$= 5702 \text{ lbf}$$

The allowable load in shear is the smallest allowable load. Therefore, masonry crushing controls.

The allowable load in tension is the lesser of the axial tensile loads governed by masonry breakout, $B_{a,b}$, and by steel yielding, $B_{a,s}$.

$$B_{a,b} = 1.25 A_{p,t}\sqrt{f'_m} \quad [\text{MSJC Eq. 2-1}]$$
$$= (1.25)(113 \text{ in}^2)\sqrt{1500 \ \frac{\text{lbf}}{\text{in}^2}}$$
$$= 5471 \text{ lbf} \quad [\text{controls}]$$

$$B_{a,s} = 0.6 A_b f_y \quad [\text{MSJC Eq. 2-2}]$$
$$= (0.6)(0.44 \text{ in}^2)\left(36{,}000 \ \frac{\text{lbf}}{\text{in}^2}\right)$$
$$= 9504 \text{ lbf}$$

Masonry breakout controls.

Check that combined shear and tension is less than or equal to 1 using the interaction equation, MSJC Eq. 2-10.

$$\frac{b_a}{B_{a,b}} + \frac{b_v}{B_{v,c}} = \frac{685 \text{ lbf}}{5471 \text{ lbf}} + \frac{822 \text{ lbf}}{1774 \text{ lbf}}$$
$$= 0.589 \quad [\leq 1]$$

However, this result is understressed, and the problem asks for the smallest bolt size. Try a $3/8$ in diameter bolt, which has an area, A_b, of 0.11 in^2.

Considering critical cases only, the allowable shear load is the lesser of

$$B_{v,c} = 350 \sqrt[4]{f'_m A_b} = 350 \sqrt[4]{\left(1500 \ \frac{\text{lbf}}{\text{in}^2}\right)(0.11 \text{ in}^2)}$$
$$= 1254 \text{ lbf} \quad [\text{controls}]$$

$$B_{v,s} = 0.36 A_b f_y = (0.36)(0.11 \text{ in}^2)\left(36{,}000 \ \frac{\text{lbf}}{\text{in}^2}\right)$$
$$= 1426 \text{ lbf}$$

Masonry crushing controls.

The allowable load in tension is the lesser of

$$B_{a,b} = 1.25 A_{p,t}\sqrt{f'_m} = (1.25)(113 \text{ in}^2)\sqrt{1500 \ \frac{\text{lbf}}{\text{in}^2}}$$
$$= 5471 \text{ lbf}$$

$$B_{a,s} = 0.6 A_b f_y = (0.6)(0.11 \text{ in}^2)\left(36{,}000 \ \frac{\text{lbf}}{\text{in}^2}\right)$$
$$= 2376 \text{ lbf} \quad [\text{controls}]$$

Steel yielding controls.

Check that the combined shear and tension is less than or equal to 1 using the interaction equation, MSJC Eq. 2-10.

$$\frac{b_a}{B_{a,s}} + \frac{b_v}{B_{v,c}} = \frac{685 \text{ lbf}}{2376 \text{ lbf}} + \frac{822 \text{ lbf}}{1254 \text{ lbf}}$$
$$= 0.944 \quad [\leq 1, \text{ OK}]$$

Since this result is near the combined stress limit, use $3/8$ in diameter bolts.

The answer is (B).

Why Other Options Are Wrong

(A) This incorrect solution mistakes the load *from* the joist as the total load *on* the joist and divides the load in half to get the resultant. It is also incorrect in only considering the shear load.

(C) This incorrect solution solves the problem using the 2005 *MSJC* provisions.

(D) This incorrect solution calculates the stresses correctly but stops with a $3/4$ in diameter bolt and does not find the smallest size bolt.

20. From a foundation design reference, the minimum depth of penetration, $D_{\min}$, for the sheet piling shown is given by the equation

$$D_{\min}^4 - \left(\frac{8H}{\gamma(k_p - k_a)b}\right) D_{\min}^2 - \left(\frac{12HL}{\gamma(k_p - k_a)b}\right)$$
$$\times D_{\min} - \left(\frac{2H}{\gamma(k_p - k_a)b}\right)^2 = 0$$

$$k_a = \tan^2\left(45° - \frac{\phi}{2}\right)$$
$$= \tan^2\left(45° - \frac{30°}{2}\right)$$
$$= 0.333$$

$$k_p = \tan^2\left(45° + \frac{\phi}{2}\right)$$
$$= \tan^2\left(45° + \frac{30°}{2}\right)$$
$$= 3.00$$

$$\gamma = 110 \text{ lbf/ft}^3$$

From the problem statement, the width of the sheet piling, b, is 2 ft; the single concentrated load, H, is

10 kips (10,000 lbf); and the length of the sheet piling above grade, L, is 10 ft.

$$D_{\min}^4 - \left(\frac{(8)(10{,}000 \text{ lbf})}{\left(110 \frac{\text{lbf}}{\text{ft}^3}\right)(3.00 - 0.333)(2 \text{ ft})}\right) D_{\min}^2$$
$$- \left(\frac{(12)(10{,}000 \text{ lbf})(10 \text{ ft})}{\left(110 \frac{\text{lbf}}{\text{ft}^3}\right)(3.00 - 0.333)(2 \text{ ft})}\right) D_{\min}$$
$$- \left(\frac{(2)(10{,}000 \text{ lbf})}{\left(110 \frac{\text{lbf}}{\text{ft}^3}\right)(3.00 - 0.333)(2 \text{ ft})}\right)^2$$
$$= 0$$

$$D_{\min}^4 - (136.3 \text{ ft}^2) D_{\min}^2 - (2045.2 \text{ ft}^3) D_{\min}$$
$$- 1161.9 \text{ ft}^4 = 0$$

This equation can be solved iteratively.

Try $D_{\min} = 10$ ft.

$$(10 \text{ ft})^4 - (136.3 \text{ ft}^2)(10 \text{ ft})^2 - (2045.2 \text{ ft}^3)(10 \text{ ft})$$
$$- 1161.9 \text{ ft}^4 = 0$$
$$-25{,}244 \text{ ft}^4 \neq 0$$

Try $D_{\min} = 15$ ft.

$$(15 \text{ ft})^4 - (136.3 \text{ ft}^2)(15 \text{ ft})^2 - (2045.2 \text{ ft}^3)(15 \text{ ft})$$
$$- 1161.9 \text{ ft}^4 = 0$$
$$-11{,}882 \text{ ft}^4 \neq 0$$

Try $D_{\min} = 16$ ft.

$$(16 \text{ ft})^4 - (136.3 \text{ ft}^2)(16 \text{ ft})^2 - (2045.2 \text{ ft}^3)(16 \text{ ft})$$
$$- 1161.9 \text{ ft}^4 = 0$$
$$-3242 \text{ ft}^4 \neq 0$$

Try $D_{\min} = 16.3$ ft.

$$(16.3 \text{ ft})^4 - (136.3 \text{ ft}^2)(16.3 \text{ ft})^2 - (2045.2 \text{ ft}^3)(16.3 \text{ ft})$$
$$- 1161.9 \text{ ft}^4 = 0$$
$$-121 \text{ ft}^4 \approx 0$$

This last solution is close to equaling zero and is precise enough for this example. Further iterations would yield an exact solution.

The minimum depth of penetration is

$$D_{\min} = 16.3 \text{ ft}$$

Applying the factor of safety of 1.3 yields

$$D = (1.3)(16.3 \text{ ft}) = 21.2 \text{ ft}$$

The total length of the sheet piling is

$$L_{\text{total}} = L + D = 10 \text{ ft} + 21.2 \text{ ft}$$
$$= 31.2 \text{ ft} \quad (31 \text{ ft})$$

The answer is (C).

Why Other Options Are Wrong

(A) This incorrect solution calculates the depth of penetration only, not the total length of the pile.

(B) This incorrect solution does not apply the factor of safety when calculating the required depth.

(D) This incorrect solution does not include the width of the sheet piling, b, in the equation for calculating the minimum depth of penetration, $D_{\min}$.

21. ASCE7 Sec. 12.12.1 specifies that the design story drift, Δ, shall not exceed the allowable story drift, Δ_a, calculated from the equations given in ASCE7 Table 12.12-1. Since the structure is rectangular in plan, torsional effects are likely insignificant and do not need to be included in the maximum story drift calculation. Therefore, the design story drift and allowable story drift will be equivalent.

Determine the appropriate allowable story drift equation from ASCE7 Table 12.12-1. From ASCE7 Table 1-1, this structure is considered an occupancy category II structure. The interior partitions, ceilings, and walls have not been designed to accommodate story drifts, and the building is not a masonry structure. Therefore, use the "all other structures" row in ASCE7 Table 12.12-1.

The story height, h_{sx}, is the height below level x (i.e., the second story), which is equivalent to the floor-to-floor height of 12 ft. The maximum design story drift at the second story is

$$\Delta = 0.020 h_{sx} = (0.020)(12 \text{ ft})\left(12 \frac{\text{in}}{\text{ft}}\right)$$
$$= 2.88 \text{ in} \quad (2.9 \text{ in})$$

The answer is (C).

Why Other Options Are Wrong

(A) This incorrect solution does not convert the story drift from feet to inches.

(B) This incorrect solution uses the allowable story drift equation for an occupancy category III structure.

(D) This incorrect solution uses the allowable story drift equation for structures with interior partitions, ceilings, and walls designed to accommodate story drifts.

22. Special reinforced masonry shear walls must comply with *MSJC* Sec. 1.17.3.2.6, Sec. 1.17.3.2.3.1, and either Sec. 2.3 or Sec. 3.3. Since is it assumed that all reinforcement is adequate to resist the applied shear loads, the provisions of *MSJC* Sec. 2.3 and Sec. 3.3 do not need to be checked.

For masonry laid in running bond, *MSJC* Sec. 1.17.3.2.6 limits the maximum spacing of vertical and horizontal reinforcement to the smallest of $L/3$, $H/3$, or 48 in.

$$s_{max} = \frac{L}{3} = \frac{(20 \text{ ft})\left(12 \frac{\text{in}}{\text{ft}}\right)}{3} = 80 \text{ in}$$

$$s_{max} = \frac{H}{3} = \frac{(10 \text{ ft})\left(12 \frac{\text{in}}{\text{ft}}\right)}{3} = 40 \text{ in} \quad [\text{controls}]$$

$$s_{max} = 48 \text{ in}$$

MSJC Sec. 1.17.3.2.6.(c).1 specifies that the minimum cross-sectional area of reinforcement in each direction must be at least 0.0007 times the gross cross-sectional area of the wall, A_g, calculated using specified dimensions. The specified thickness of a nominal 12 in concrete masonry wall is 11.63 in.

$$A_{v,min} = 0.0007 A_g = (0.0007)(11.63 \text{ in})(20 \text{ ft})\left(12 \frac{\text{in}}{\text{ft}}\right)$$
$$= 1.95 \text{ in}^2$$

$$A_{h,min} = 0.0007 A_g = (0.0007)(11.63 \text{ in})(10 \text{ ft})\left(12 \frac{\text{in}}{\text{ft}}\right)$$
$$= 0.977 \text{ in}^2$$

MSJC Sec. 1.17.3.2.3.1 provides the minimum reinforcement requirements for corners, openings, and ends of walls. *MSJC* specifies that the minimum horizontal reinforcement must be at least 0.2 in^2 of bond beam reinforcement spaced no more than 120 in on center vertically, or two W1.7 joint reinforcing wires with a maximum spacing of 16 in on center vertically.

Determine the minimum required horizontal reinforcement. Try two no. 4 bars in bond beams spaced at 48 in on center vertically.

$$A_h = \frac{(2)(0.20 \text{ in}^2)(10 \text{ ft})\left(12 \frac{\text{in}}{\text{ft}}\right)}{48 \text{ in}}$$
$$= 1.00 \text{ in}^2 \quad [> A_{h,min}, \text{ OK}]$$

From *MSJC* Sec. 1.17.3.2.6, the maximum spacing is limited to 40 in. Try one no. 5 bar in bond beams spaced at 32 in on center vertically.

$$A_h = \frac{(1)(0.31 \text{ in}^2)(10 \text{ ft})\left(12 \frac{\text{in}}{\text{ft}}\right)}{32 \text{ in}}$$
$$= 1.16 \text{ in}^2 \quad [> A_{h,min}, \text{ OK}]$$

Use one no. 5 bar in bond beams spaced at 32 in on center vertically.

Determine the minimum required vertical reinforcement. Try no. 5 reinforcing bars spaced at 24 in on center.

$$A_v = \frac{(0.31 \text{ in}^2)(20 \text{ ft})\left(12 \frac{\text{in}}{\text{ft}}\right)}{24 \text{ in}}$$
$$= 3.10 \text{ in}^2 \quad [> A_{v,min}, \text{ OK}]$$

Because this is more than adequate for the minimum requirement, try no. 5 reinforcing bars spaced at 32 in on center.

$$A_v = \frac{(0.31 \text{ in}^2)(20 \text{ ft})\left(12 \frac{\text{in}}{\text{ft}}\right)}{32 \text{ in}}$$
$$= 2.33 \text{ in}^2 \quad [> A_{v,min}, \text{ OK}]$$

Therefore, use no. 5 vertical reinforcing bars spaced at 32 in on center.

Check the requirements of *MSJC* Sec. 1.17.3.2.6.(c).

$$A_v + A_h \geq 0.002 A_g$$

$$0.002 A_g = (0.002)(11.63 \text{ in})(10 \text{ ft})\left(12 \frac{\text{in}}{\text{ft}}\right) = 2.79 \text{ in}^2$$

$$A_v + A_h = 2.33 \text{ in}^2 + 1.16 \text{ in}^2$$
$$= 3.49 \text{ in}^2 \quad [\geq 0.002 A_g, \text{ OK}]$$

The answer is (B).

Why Other Options Are Wrong

(A) This incorrect solution only calculates the minimum horizontal reinforcement. Minimum shear reinforcement is required in both directions according to *MSJC* Sec. 1.17.3.2.6.

(C) This solution incorrectly omits the check for the maximum horizontal spacing required by *MSJC* Sec. 1.17.3.2.6.

(D) This incorrect solution bases the minimum required horizontal and vertical reinforcement on $0.002 A_g$. This is the minimum required for the sum of the horizontal and vertical reinforcements, not the individual amounts.

23. ASCE7 Chap. 14 contains the material-specific requirements for seismic detailing. ASCE7 Sec. 14.1 references AISC 341, which lists seismic requirements for structural steel buildings. AISC 341 Sec. 9 covers SMF requirements, and Sec. 9.2a states that the beam-to-column connections used in a seismic load-resisting system shall be capable of sustaining an interstory drift angle of at least 0.04 rad (not 0.02 rad, as in option A). An interstory drift angle of at least 0.02 rad is a requirement for intermediate moment frames [AISC 341 Sec. 10.2a].

The answer is (A).

Why Other Options Are Wrong

(B) AISC 341 Sec. 9.2b lists the use of beam-to-column connections prequalified for SMF in accordance with AISC 341 App. P to satisfy the requirements of AISC 341 Sec. 9.2a.

(C) AISC 341 Sec. 9.8 requires that both flanges of beams be laterally braced.

(D) AISC 341 Sec. 9.8 specifies that lateral bracing of beam flanges have a maximum spacing of $0.086 r_y E/F_y$.

24. For shear walls sheathed with different (dissimilar) materials on opposite sides of the wall, SDPWS Sec. 4.3.3.3.2 specifies that the combined nominal unit seismic shear capacity must be the greater of (a) two times the smaller nominal unit shear capacity, $v_{s,\text{min}}$, or (b) the larger nominal unit shear capacity, $v_{s,\text{max}}$.

SDPWS Table 4.3A gives the tabulated nominal shear capacities for seismic and wind design of a wood-frame shear wall sheathed with wood-based panels. For $3/8$ in wood structural panel-sheathing (OSB) with 8d nails and a panel edge spacing of 6 in, the nominal unit seismic shear capacity, v_s, is 440 lbf/linear ft.

From SDPWS Table 4.3C, for a wood-frame shear wall sheathed with $1/2$ in gypsum wallboard, attached with no. 6 type S drywall screws spaced at 8 in, and having unblocked studs spaced at 16 in on center, the nominal unit seismic shear capacity, v_s, is 120 lbf/linear ft.

The combined nominal unit seismic shear capacity is the larger of

$$v_{sc} = 2v_{s,\text{min}} = (2)\left(120 \ \frac{\text{lbf}}{\text{linear ft}}\right) = 240 \ \text{lbf/linear ft}$$
$$v_{sc} = v_{s,\text{max}} = 440 \ \text{lbf/linear ft} \quad [\text{controls}]$$

The answer is (B).

Why Other Options Are Wrong

(A) This incorrect solution uses the smaller of the two combined capacities rather than the larger.

(C) This incorrect solution sums the shear capacities for each side rather than following the provisions of SDPWS Sec. 4.3.3.3.2.

(D) This incorrect solution uses the nominal unit seismic shear capacity for a blocked wall and sums the shear capacities for each side.

25. According to SDPWS Sec. 4.3.7.5, wood-framed shear walls sheathed with gypsum wallboard are permitted to resist seismic forces in seismic design categories A through D. Statement I is true.

From SDPWS Table 4.3C, wood-framed shear walls sheathed with gypsum wallboard can be constructed as blocked or unblocked walls. Statement II is true.

SDPWS Sec. 4.3.7.3 covers particleboard shear walls and permits their use only in seismic design categories A, B, and C. Statement III is false.

SDPWS Sec. 4.3.7.6 states that single-layer lumber used to diagonally sheathe wood-frame shear walls must have a nominal thickness of at least 1 in. Statement IV is true.

The answer is (C).

Why Other Options Are Wrong

(A) This incorrect solution correctly identifies statement I as true, but statements II and IV are also true.

(B) This incorrect solution correctly identifies statement II as true, but statement III is false.

(D) This incorrect solution correctly identifies statements II and IV as true, but statement III is false.

26. From AASHTO Table 3.7.3.1-1, the drag coefficient, C_D, on a semicircular nosed pier is 0.7. Using AASHTO Sec. 3.7.3.1 and Eq. 3.7.3.1-1, determine the longitudinal drag pressure.

$$P = \frac{C_D v^2}{1000} = \frac{(0.7)\left(10 \ \frac{\text{ft}}{\text{sec}}\right)^2}{1000} = 0.07 \ \text{kip/ft}^2$$

The longitudinal drag force is

$$F = PA = \left(0.07 \ \frac{\text{kip}}{\text{ft}^2}\right)(10 \ \text{ft})(5 \ \text{ft}) = 3.5 \ \text{kips}$$

The answer is (B).

Why Other Options Are Wrong

(A) This incorrect solution did not square the velocity in the drag pressure calculation.

(C) This incorrect solution assumes flow in the wrong direction and uses a drag coefficient of 1.4.

(D) This solution incorrectly uses the length of the pier (30 ft) instead of the width of the pier (5 ft) when determining the drag force.

27. ASCE7 Sec. 12.11 contains the requirements for structural walls and their anchorage. According to ASCE7 Sec. 12.11.2, a wall must be capable of resisting the greatest of

$$F = 0.4 S_{DS} I W_{wall} \quad [\text{ASCE7 Sec. 12.11.1}]$$
$$F = 0.10 W_{wall} \quad [\text{ASCE7 Sec. 12.11.1}]$$
$$F_{lbf/linear\,ft} = 400 S_{DS} I \quad [\text{ASCE7 Sec. 12.11.2}]$$
$$F = 280 \text{ lbf/linear ft} \quad [\text{ASCE7 Sec. 12.11.2}]$$

The importance factor, I, is based on the occupancy category. ASCE7 Table 1-1 indicates that a large elementary school is in occupancy category III. ASCE7 Table 11.5-1 assigns an importance factor of 1.25 to all occupancy category III buildings.

The force that the anchorage must resist is the greatest of

$$F = 0.4 S_{DS} I W_{wall}$$
$$= (0.4)(0.46)(1.25)\left(1350 \; \frac{\text{lbf}}{\text{linear ft}}\right)$$
$$= 310.5 \text{ lbf/linear ft} \quad (311 \text{ lbf/linear ft})$$

[controls]

$$F = 0.10 W_{wall} = (0.10)\left(1350 \; \frac{\text{lbf}}{\text{linear ft}}\right)$$
$$= 135 \text{ lbf/linear ft}$$
$$F_{lbf/linear\,ft} = 400 S_{DS} I = (400)(0.46)(1.25)$$
$$= 230 \text{ lbf/linear ft}$$
$$F = 280 \text{ lbf/linear ft}$$

The answer is (C).

Why Other Options Are Wrong

(A) This incorrect solution uses an importance factor of 1.0 instead of 1.25.

(B) This incorrect solution satisfies the minimum anchorage force requirements found in ASCE7 Sec. 12.11.2, ignoring those given in Sec. 12.11.1.

(D) This incorrect solution adds the weight of the plank floor to the wall weight, ignoring the difference in units.

28. IBC Sec. 1710.2 provides the structural observation requirements for seismic resistance. For seismic design category D, structural observations are required if the structure is classified as occupancy category III or IV or if the structure height is greater than 75 ft. From IBC Table 1604.5, a high school is an occupancy category III structure.

The answer is (C).

Why Other Options Are Wrong

(A) An office building is classified as an occupancy category II structure. Only occupancy category II structures assigned to seismic design category E that are greater than two stories above grade require structural observations.

(B) An office building is classified as an occupancy category II structure. Though this structure is more than two stories, only occupancy category II structures assigned to seismic design category E that are greater than two stories above grade require structural observations.

(D) An agricultural building is classified as an occupancy category I structure, and this structure has a height less than 75 ft.

29. OSHA Std. 1904 Subpart B and Subpart C specify the scope and criteria of record keeping and forms for fatalities and injuries.

OSHA Std. 1904.29, item (a) states that employers must use OSHA forms 300 ("Log of Work-Related Injuries and Illnesses"), 300A ("Summary of Work-Related Injuries and Illnesses"), and 301 ("Injury and Illness Incident Report") to record workplace injuries and illnesses.

OSHA Std. 1904.1, item (a) states that if an employer has ten or fewer employees at all times during a calendar year, OSHA safety records do not need to be kept.

OSHA Std. 1904.5 covers accidents that are not considered work-related. OSHA Table 1904.5(b)(2) specifies that accidents are not considered work-related if "the injury or illness is caused by a motor vehicle accident and occurs on a company parking lot or company access road while the employee is commuting to or from work."

OSHA Std. 1904.39, item (a) states that within 8 hr after the death of any employee from a work-related incident or the in-patient hospitalization of three or more employees from a work-related incident, an employer must orally report the incident(s) to OSHA offices via phone or in person.

The answer is (C).

Why Other Options Are Wrong

(A) This incorrect option represented a true statement per OSHA regulations.

(B) This incorrect option represented a true statement per OSHA regulations.

(D) This incorrect option represented a true statement per OSHA regulations.

30. ACI 318 Sec. 7.4.1 states that when concrete is placed, reinforcements must be free from mud, oil, or other nonmetallic coatings that can decrease bonding. Thus, the mud must be cleaned (e.g., sand blasted) from the coated reinforcements.

ACI 318 Sec. 7.4.2 states that except for prestressing steel, steel reinforcement with rust, mill scale, or a combination of both is generally satisfactory. (The minimum dimensions, including height of deformations, and the weight of a test specimen wire-brushed by hand must comply with applicable ASTM specifications referenced in ACI 318 Sec. 3.5.) In this case, since the mill scale is minor, it should be acceptable according to ACI 318 Sec. 7.4.2.

The answer is (C).

Why Other Options Are Wrong

(A) This incorrect option unnecessarily rejects the entire shipment.

(B) This incorrect option unnecessarily rejects a portion of the shipment.

(D) This incorrect option neglects the requirements of ACI 318 Sec. 7.4.1.

Trust PPI for Your Structural SE Exam Review Needs
For more information, visit www.ppi2pass.com.

Comprehensive Reference and Practice Materials

Structural Engineering Reference Manual
Alan Williams, PhD, SE, FICE, C Eng

- A complete introduction to the exam format and content
- Full coverage of exam topics and relevant codes
- More than 700 equations
- Over 270 practice problems with complete solutions
- Key tables, charts, and figures

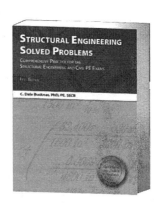

Structural Engineering Solved Problems
C. Dale Buckner, PhD, PE, SECB

- Comprehensive practice for the
  – Structural SE exam
  – Civil PE exam's structural depth section
- 100 practice problems arranged in order of increasing complexity
- Complete step-by-step solutions

Timber Design for the Civil and Structural PE Exams
Robert H. Kim, MSCE, PE;
Jai B. Kim, PhD, PE; with
Parker E. Terril, BSCE, MSBE

- A complete overview of the relevant codes and standards
- 40 design examples
- 6 scenario-based practice problems
- Easy-to-use tables, figures, and timber design nomenclature

Concrete Design for the Civil and Structural PE Exams
C. Dale Buckner, PhD, PE

- Comprehensive concrete design review for the Structural SE and Civil PE exams
- A complete overview of relevant codes and standards
- 37 practice problems
- Easy-to-use tables, figures, and concrete design nomenclature

Steel Design for the Civil PE and Structural SE Exams
Frederick S. Roland, PE, SECB, RA, CFEI, CFII

- Comprehensive overview of the key elements of structural steel design and analysis
- Side by side LRFD and ASD solutions
- More than 50 examples and 35 practice problems
- Solutions cite specific *Steel Manual* sections, equations, or table numbers

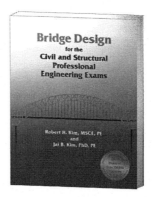

Bridge Design for the Civil and Structural Professional Engineering Exams
Robert H. Kim, MSCE, PE; and
Jai B. Kim, PhD, PE

- An overview of key bridge design principles
- 5 design examples
- 2 practice problems
- Step-by-step solutions

Don't miss all the Structural SE exam news, the latest exam advice, the exam FAQs, and the unique community of the Exam Forum at **www.ppi2pass.com**.